图书在版编目(CIP)数据

北林国际花园建造节 = BFU INTERNATIONAL GARDEN-MAKING FESTIVAL / 王向荣等编著. —北京：中国建筑工业出版社，2021.9

ISBN 978-7-112-26419-3

Ⅰ. ①北… Ⅱ. ①王… Ⅲ. ①园林艺术－展览会－概况－北京 Ⅳ. ①TU986.1-282.1

中国版本图书馆CIP数据核字（2021）第149654号

责任编辑：杜 洁 兰丽婷
责任校对：王 烨

北林国际花园建造节
BFU INTERNATIONAL GARDEN-MAKING FESTIVAL
王向荣 赵晶 郑曦 等 编著
WANG Xiangrong ZHAO Jing ZHENG Xi, etc.

*

中国建筑工业出版社出版、发行（北京海淀三里河路9号）
各地新华书店、建筑书店经销
北京富诚彩色印刷有限公司印刷

*

开本：889毫米×1194毫米 1/24 印张：$15^1/_3$ 字数：253千字
2021年9月第一版 2021年9月第一次印刷
定价：149.00元
ISBN 978-7-112-26419-3
(37963)

北林国际花园建造节

BFU INTERNATIONAL GARDEN-MAKING FESTIVAL

王向荣　赵晶　郑曦 等　编著

WANG Xiangrong　ZHAO Jing　ZHENG Xi, etc.

中国建筑工业出版社

CHINA ARCHITECTURE & BUILDING PRESS

参与人员（以姓名拼音首字母排序）

2018 年：

曹 娟　崔庆伟　董 璁　董 丽　段 威　范舒欣　戈晓宇　郭 巍　郝培尧　何禹璇　匡 纬　李方正
李冠衡　李 倞　李燕妮　李 正　刘宝昌　刘 伟　刘 尧　刘昱霏　刘志成　林辰松　马晓艳　彭 腾
钱 云　尚 书　宋 文　王博娅　王丹丹　王沛永　王思元　王向荣　吴丹子　魏 芳　韦诗誉　徐 桐
许晓明　肖 遥　尹 豪　杨晓东　严亚瓴　周春光　赵 晶　张诗阳　张 伟　郑 曦　郑小东　张云路

2019 年：

曹 娟　崔庆伟　董 璁　董 丽　段 威　范舒欣　郭 巍　戈晓宇　匡 纬　李方正　李 正　刘 通
刘 伟　刘 尧　刘昱霏　刘志成　林辰松　马 嘉　彭 腾　王博雅　王丹丹　王思元　王向荣　吴丹子
魏 方　韦诗誉　肖 遥　杨晓东　尹 豪　张诗阳　赵 晶　郑 曦　郑小东　周春光

2020 年：

边思敏　陈明坤　董 丽　段 威　冯 黎　郝培尧　康 瑛　李方正　李见哲　李梦姣　刘昱霏　祁 军
钱 云　荣 科　申静霞　王向荣　王晞月　王永林　王韵双　翁嘉蔚　吴本虹　吴玥瑶　谢玉常　杨小广
张 彤　张诗阳　张沚晴　赵 晶　郑 曦

Participants (Sort by Initials)

2018:

Cao Juan, Cui Qingwei, Dong Cong, Dong Li, Duan Wei, Fan Shuxin, Ge Xiaoyu, Guo Wei, Hao Peiyao, He Yuxuan, Kuang Wei, Li Fangzheng, Li Guanheng, Li Jing, Li Yanni, Li Zheng, Liu Baochang, Liu Wei, Liu Yao, Liu Yufei, Liu Zhicheng, Lin Chensong, Ma Xiaoyan, Peng Teng, Qian Yun, Shang Shu, Song Wen, Wang Boya, Wang Dandan, Wang Peiyong, Wang Siyuan, Wang Xiangrong, Wu Danzi, Wei Fang, Wei Shiyu, Xu Tong, Xu Xiaoming, Xiao Yao, Yin Hao, Yang Xiaodong, Yan Yaling, Zhou Chunguang, Zhao Jing, Zhang Shiyang, Zhang Wei, Zheng Xi, Zheng Xiaodong, Zhang Yunlu

2019:

Cao Juan, Cui Qingwei, Dong Cong, Dong Li, Duan Wei, Fan Shuxin, Guo Wei, Ge Xiaoyu, Kuang Wei, Li Fangzheng, Li Zheng, Liu Tong, Liu Wei, Liu Yao, Liu Yufei, Liu Zhicheng, Lin Chensong, Ma Jia, Peng Teng, Wang Boya, Wang Dandan, Wang Siyuan, Wang Xiangrong, Wu Danzi, Wei Fang, Wei Shiyu, Xiao Yao, Yang Xiaodong, Yin Hao, Zhang Shiyang, Zhao Jing, Zheng Xi, Zheng Xiaodong, Zhou Chunguang

2020:

Bian Simin, Chen Mingkun, Dong Li, Duan Wei, Feng Li, Hao Peiyao, Kang Ying, Li Fangzheng, Li Jianzhe, Li Mengjiao, Liu Yufei, Qi Jun, Qian Yun, Rong Ke, Shen Jingxia, Wang Xiangrong, Wang Xiyue, Wang Yonglin, Wang Yunshuang, Weng Jiawei, Wu Benhong, Wu Yueyao, Xie Yuchang, Yang Xiaoguang, Zhang Tong, Zhang Shiyang, Zhang Zhiqing, Zhao Jing, Zheng Xi

花 园

The Flower Garden

贝尔托·布莱希特

Bertolt Brecht

湖畔

By the lake

在冷杉和银白杨林中

In the fir and silver aspen forest

被墙和灌木丛护卫着 有一个花园

There is a garden guarded by walls and bushes

巧妙地种植着

Cleverly planted

不同季节的花卉

Flowers in different seasons

每年从三月到十月

There are flowers in full bloom here

这里都有鲜花盛开

Every year from March to October

清晨

In early morning

有时候坐在这里

Sometimes I sit here

期望我也会这样

Expect myself to do the same

无论什么时候 不管天气好坏

No matter the weather is good or bad

都能拿出某些个

I can take out something

让人喜欢的东西

That makes people like it

目　录
CONTENTS

北林国际花园建造节介绍

Introduction of BFU International Garden-Making Festival

参与建造院校 PARTICIPATING INSTITUTIONS

北京林业大学
Beijing Forestry University

广州美术学院
Guangzhou Academy of Fine Arts

华中科技大学
Huazhong University of Science and Technology

北京建筑大学
Beijing University of Civil Engineering and Architecture

哈尔滨工业大学
Harbin Institute of Technology

南京林业大学
Nanjing Forestry University

成都理工大学
Chengdu University of Technology

墨尔本皇家理工大学
Royal Melbourne Institute of Technology, Australia

南京师范大学中北学院
Nanjing Normal University Zhongbei College

重庆大学
Chongqing University

韩国庆熙大学
Kyung Hee University, Korea

清华大学
Tsinghua University

重庆交通大学
Chongqing Jiaotong University

华南理工大学
South China University of Technology

日本国立千叶大学
Chiba University, Japan

东南大学
Southeast University

华南农业大学
South China Agricultural University

四川大学
Sichuan University

四川农业大学
Sichuan Agricultural University

西安建筑科技大学
Xi 'an University of Architecture and Technology

香港珠海学院
Chu Hai College of Higher Education, Hong Kong

山东建筑大学
Shandong Jianzhu University

西北农林科技大学
Northwest A & F University

西南林业大学
Southwest Forestry University

上海交通大学
Shanghai Jiao Tong University

西华大学
Xihua University

浙江农林大学
Zhejiang A & F University

天津大学
Tianjin University

西南大学
Southwest University

中央美术学院
Central Academy of Fine Arts

泰国朱拉隆功大学
Chulalongkorn University, Thailand

西南交通大学
Southwest Jiaotong University

同济大学
Tongji University

西南民族大学
Southwest Minzu University

2018 评委简介 JUDGES

王向荣
北京林业大学园林学院院长、教授
WANG Xiangrong,
Dean and professor of the School of Landscape Architecture, Beijing Forestry University

Ron Henderson
美国伊利诺伊理工学院风景园林、城市规划系主任、教授
Ron Henderson,
Director of the Landscape Architecture+Urbanism Program, Illinois Institute of Technology

Simon Bell
英国爱丁堡艺术学院教授
Simon Bell,
Professor of the Edinburgh College of Art, UK

Robert Ryan
美国马萨诸塞大学阿默斯特分校风景园林与区域规划系主任、教授
Robert Ryan,
Department Chair and Professor of Landscape Architecture and Regional Planning, University of Massachusetts at Amherst

Peter C. Bosselmann
美国加利福尼亚大学伯克利分校建筑、城市与区域规划、景观建筑与城市设计系荣誉退休教授
Peter C. Bosselmann,
Professor Emeritus of the Graduate School in Architecture, City & Regional Planning, Landscape Architecture and Urban Design, University of California at Berkeley

Kim, Jin-Oh
韩国庆熙大学艺术与设计学院风景园林系主任
Kim, Jin-Oh,
Chair, Department of Landscape Architecture, College of Art & Design, Kyung Hee University

Ilke Marschall
德国爱尔福特应用科技大学园林学院教授
Ilke Marschall,
Professor of the Department of Landscape Architecture, University of Applied Sciences Erfurt, Germany

Louise A. Mozingo
美国加利福尼亚大学伯克利分校环境设计学院风景园林与环境规划教授
Louise A. Mozingo,
Professor of Landscape Architecture & Environmental Planning, College of Environmental Design, University of California at Berkeley

Dirk Sijmons
荷兰代尔夫特大学风景园林系前系主任、教授
Dirk Sijmons,
Former Chair and Professor of Landscape Architecture, Delft University of Technology

Rik de Visser
荷兰 VISTA 景观设计与城市规划总监
Rik de Visser,
Director of Landscape Design and Urban planning, Vista

下村彰男
日本东京大学农学生命科学研究科教授
SHIMOMURA Akio,
Professor of the Graduate School of Agricultural and Life Sciences, University of Tokyo

蔡 卫
浙江竹境文化旅游发展股份有限公司董事长
CAI Wei,
Chairman of Zhejiang Bamboo Cultural Tourism Development Co., Ltd

李雄
北京林业大学副校长、园林学院教授
LI Xiong,
Vice President of Beijing Forestry University, Professor of the School of Landscape Architecture

宋晔皓
清华大学建筑学院教授
SONG Yehao,
Professor of the School of Architecture, Tsinghua University

于学斌
北京市园林绿化集团副总经理、北京市花木有限公司董事长、总经理
YU Xuebin,
Deputy General Manager of Beijing Landscape and Forestry Group Co., Ltd./ Chairman of General Manager Beijing Florascape Co., Ltd.

赵万民
重庆大学建筑与城市规划学院原院长、博士生导师、教授
ZHAO Wanmin,
Former Dean, Doctoral Supervisor/ Professor, School of Architecture and Urban Planning, Chongqing University

2019 评委简介 JUDGES

王向荣
北京林业大学园林学院院长、教授
WANG Xiangrong,
Dean and professor of the School of Landscape Architecture, Beijing Forestry University

Ed Wall
英国格林尼治大学风景园林学院院长
Ed Wall,
Dean of the School of Landscape Architecture, University of Greenwich, UK

Bradley Cantrell
美国弗吉尼亚大学风景园林系主任、教授
Bradley Cantrell,
Director and professor of the Department of Landscape Architecture, University of Virginia, USA

Elizabeth Mossop
澳大利亚悉尼科技大学设计建筑与建筑系主任、教授
Elizabeth Mossop,
Professor and director of the Department of Design Architecture and Architecture, University of Technology, Sydney, Australia

蔡 卫
浙江竹境文化旅游发展股份有限公司董事长
CAI Wei,
Chairman of Zhejiang Bamboo Cultural Tourism Development Co., Ltd

James Hayter
国际风景园林师联合会（IFLA）主席
James Hayter,
Chairman of International Federation of Landscape Architects

David Biggs
美国加利福尼亚大学河滨分校历史地理副教授
David Biggs,
Associate Professor of history and geography in the University of California, Riverside

Jose Alfredo Ramirez Galindo
英国建筑联盟学院 March/MSC 研究生课程的联合发起人
Jose Alfredo Ramirez Galindo,
Co-sponsor of the March/MSC postgraduate courses, director and co-director of the Ground Laboratory, Architectural Association School of Architecture. UK

Kim, Jin-Oh
韩国庆熙大学艺术与设计学院风景园林系主任
Kim, Jin-Oh,
Chair, Department of Landscape Architecture, College of Art & Design, Kyung Hee University

Steffen Nijhuis
荷兰代尔夫特理工大学建筑与建成环境学院城市系副教授
Steffen Nijhuis,
Associate professor of the Department of Urbanism, School of Architecture and Built-up Environment, Delft University of Technology, the Netherlands

刘可为
国际竹藤组织全球竹建筑项目协调员
LIU Kewei,
International Bamboo and Rattan Organisation Project Coordinator

Udo Weilacher
德国慕尼黑工业大学建筑学院风景园林和工业景观系教授
Udo Weilacher,
Professor of the Department of Landscape Architecture and Industrial Landscape, School of Architecture, Technical University of Munich, Germany

Makoto Yokohar
日本东京大学景观教授
Makoto Yokohar,
Professor of Landscape, University of Tokyo, Japan

Valentin Johannes (Han) Meyer
荷兰代尔夫特理工大学建筑与建成环境学院城市系教授
Valentin Johannes (Han) Meyer,
Professor of the Department of Urbanism, School of Architecture and Built-up Environment, Delft University of Technology, the Netherlands

Nicolas Godelet
NICOLAS GODELET 建筑事务所创始人、比利时工程师、建筑师
Nicolas Godelet,
Founder of NICOLAS GODELET Architects & Engineers Nicolas Godelet

2020 评委简介 JUDGES

王向荣
北京林业大学园林学院院长 / 教授
WANG Xiangrong,
Dean and professor of the School of Landscape Architecture, Beijing Forestry University

李树华
清华大学建筑学院景观学系教授，亚洲园艺疗法联盟、绿色疗法与康养景观研究中心主席
LI Shuhua,
Professor, Department of Landscape, School of Architecture, Tsinghua University

刘可为
国际竹藤组织全球竹建筑项目协调员
LIU Kewei,
International Bamboo and Rattan Organisation, Project Coordinator

陈明坤
成都市公园城市建设发展研究院院长、教授级高级工程师、注册城市规划师
CHEN Mingkun,
President of Chengdu Park City Construction and Development Research Institute, professor level senior engineer, registered urban planner

王香春
中国城市建设研究院城乡生态文明研究院院长、中国公园协会秘书长、教授级高级工程师
WANG Xiangchun,
Secretary General of China Park Association, professor level senior engineer

张彤
成都市花木技术服务中心主任、高级工程师
ZHANG Tong,
Director and senior engineer of Chengdu Flower and Wood Technology Service Center

刘滨谊
同济大学建筑与城市规划学院教授、博士生导师、风景园林学科专业委员会主任、风景科学研究所所长
LIU Binyi,
Professor and doctoral supervisor of School of architecture and urban planning, Tongji University, director of Landscape Architecture Discipline Committee and director of Institute of Landscape Science

青山周平
B.L.U.E. 建筑设计事务所创始合伙人、主持建筑师
Shuhei Aoyama,
Founding partner and chief architect of B.L.U.E. Architectural Design Office

贾建中
中国风景园林学会秘书长、教授级高级工程师
JIA Jianzhong,
Secretary general and professor level senior engineer of Chinese society of Landscape Architecture

朱育帆
清华大学建筑学院景观学系副系主任、教授
ZHU Yufan,
Associate director and Professor, Department of landscape, School of architecture, Tsinghua University

李雄
北京林业大学副校长、园林学院教授
LI Xiong,
Vice President of Beijing Forestry University, Professor of the School of Landscape Architecture

宋晔皓
清华大学建筑学院教授
SONG Yehao,
Professor of the School of Architecture, Tsinghua University

蔡 卫
浙江竹境文化旅游发展股份有限公司董事长
CAI Wei,
Chairman of Zhejiang Bamboo Cultural Tourism Development Co., Ltd

在我国，早期虽然没有园林展的形式，但是自古以来以花卉等植物为展览内容供人观赏，就是人们喜闻乐见的活动。早在魏晋南北朝时期，帝王便在宫苑内集中栽植一些品种名贵的植物。唐代产生了专门的花市，展出与售卖园圃中培育的花卉，如洛阳的牡丹花会等。至宋代，人们开始考虑花卉展览的选址、布置手法、栽植材料和品种等内容，形成了真正意义上的花卉展览。明清时期花卉展览日趋成熟，并开始注重景观的营造，植物通常配合水池、假山、建筑等以突显主题。

在西方，园林展已有上百年的历史，在组织策划、规划设计、市场运作、文化艺术交流等方面积累了丰富的经验。伴随着中西方园林文化艺术的沟通交流，园林展作为一种新鲜事物传入中国。

自 1999 年昆明世界园艺博览会起，全国各地纷纷开始举办包括国际性园林展、综合性园林园艺博览会、花博会、绿博会等各类园林展。举办园林展是绿色产业、技术和文化发展较为成熟且相互融合的重要标志。

图片来源：https://www.sohu.com/a/215413820_167180

Although there was not the form of garden exhibitions in the early days in China, showcases of arrangement of flowers and plants had been provided for people to appreciate since ancient times, which had long been an activity that people loved to take part in. As early as in the Wei, Jin and Southern and Northern Dynasties, the emperors had some rare plants centrally planted in their palaces. In the Tang Dynasty, special flower markets appeared to display and sell the flowers cultivated in gardens, such as the Peony Fair in Luoyang. In the Song Dynasty, people began to consider the site selection, layout techniques, planting materials and varieties for flower exhibitions, thus forming the flower exhibition in its true meaning. During the Ming and Qing Dynasties, flower exhibitions became more and more mature, and people began to pay attention to the creation of landscapes. Plants usually matched pools, rockeries, buildings, etc. to highlight the themes.

In the West, garden exhibitions have a history of hundreds of years, with rich experiences accumulated in organizing and devising the plans, designs, market operations, and cultural and artistic exchanges. With the communication and exchange between China and Western nations in garden culture and art, garden exhibitions were introduced to China as a new thing.

Since the Kunming World Horticultural Expo in 1999, various garden exhibitions have been held across China, including international garden exhibitions, comprehensive gardening and horticultural exhibitions, flower fairs, green fairs and the like. Garden exhibitions are an important sign for the maturity and integrated development of green industries, technologies and cultures.

其中，有一类园林展在国际上得到了较好的发展，但是在国内发展并不广泛，便是艺术性花园展。艺术性花园展是以推动园林艺术探索与发展、向公众普及园林艺术思想与应用为主要目的的一类园林展，一般以展示园林艺术思想、技术、材料的展览花园为主体，或结合其他相关的艺术形式，如雕塑、工艺产品、装置艺术等，是一种能够呈现园林艺术多样化表达、挑战行业传统理论思想、提供行业发展方向可能性的综合艺术形式。

在世界范围内，艺术性花园展的出现虽然只有几十年的时间，却发展迅猛，越来越普及，主办机构从地方政府到民间机构，选址也多会在一些公园、展览中心等，多在固定的展期面向公众开放。

1992 年，在法国的肖蒙古堡举行了世界首个艺术性花园展——法国肖蒙城堡国际花园艺术节。之后肖蒙花园节每年举办一次，由组委会提出特定的展览主题，在世界范围征集花园设计方案，同时也会邀请几位世界知名设计师，得到了社会关注和设计师的广泛参与。整个展示区面积约 $3hm^2$，分为 30 个独立的展示空间，每个空间占地约 $250m^2$。展园气氛多轻松愉悦或发人深省，使用新奇的材料或装置挑战传统花园的概念，加深了大众对花园的理解，推动了行业的发展。自此，其他国家也相继开始了对艺术性花园展的探索。

图片来源：http://www.chla.com.cn/htm/2019/0606/272145.html

However, the artistic garden exhibition, one type of garden exhibition that has been well developed internationally, is not widely developed in China. As a type of garden exhibition mainly purposed to promote the exploration and development of garden art and popularize garden art ideas and applications to the public, the artistic garden exhibitions are generally focused on the exhibition gardens that display garden art ideas, techniques and materials, or combined with other related art forms, such as sculptures, craft products, installation art, etc. Therefore, they are a comprehensive art form that can present diverse expressions of garden art, challenge the traditional theories and ideas of the industry, and provide new possibility for the industry development.

Around the world, although artistic garden exhibitions have only been developed for a few decades, they have grown rapidly and become increasingly popular. Organizers of them include local governments to private organizations. And they are often located in parks, exhibition centers, etc., open to the public during a fixed period of time.

In 1992, the world's first artistic garden exhibition was held in Chaumont sur Loire, France: the International Garden Festival in Chaumont sur Loire. Afterwards, the Chaumont Garden Festival has been held once a year, for which the organizing committee proposed a specific exhibition theme and solicit garden design proposals from around the world, while inviting several world-renowned designers. So it has attracted social attention and been extensively participated by designers. The entire exhibition region covers an area of about 3 hm^2 and is divided into 30 independent exhibition spaces, each covering an area of about 250 m^2. The atmosphere of the exhibition garden is often relaxed and pleasant or thought-provoking. Novel materials or installations are used to challenge the concept of traditional gardens, deepen the public's understanding of gardens and promote the development of the industry. Since then, other countries have also begun to explore artistic garden exhibitions.

图片来源：http://www.dylandscape.com/

加拿大国际花园节于 2000 年在加拿大梅蒂斯市的里弗德花园首次举办，此后每年举办一届。花园节主旨是重释与挖掘花园的新意义，为设计师提供实验和创新的场地。这个花园节也充分体现了花园实践的创新性和实验性，并尽可能结合视觉艺术、建筑等其他元素。展园由高大的桦树分隔为若干个 200m^2 的矩形场地，形成相对独立的花园空间。

新加坡花园节是目前世界上唯一在热带地区举办的国际花园展，于 2006 年 12 月由新加坡国家公园局在新达城国际会议与展览中心首次举办，此后每两年举办一届。2018 年第七届花园节在新加坡滨海湾花园举行，总占地面积达 2.2hm^2，共邀请 33 名海内外著名园艺设计师，并吸引了不少民间园艺团体和公众。新加坡花园节首次走出了室内展厅，增加了室外展区，让公众有机会欣赏到更具规模、更壮观的热带园艺与花艺展示。除了这些花园节外，世界各地也有不同的花园节，如英国韦斯顿伯特花园展、葡萄牙蓬蒂迪利玛国际花园节等。

图片来源：https://www.facebook.com/events/1323500871156773/

图片来源：https://www.facebook.com/sggardenfest/

The International Garden Festival at Jardins de Metis was first held in 2000 in Reford Gardens in Metis, Canada, and has been held annually thereafter. The main purpose of the Garden Festival is to reinterpret and explore the new meanings of gardens and to provide designers with a venue for experimentation and innovation. This garden festival also fully embodies the innovative and experimental features of garden practice, and combines visual art, architecture and other elements as much as possible. The exhibition garden is divided into several 200 m^2 rectangular courts with tall birch trees, forming many relatively independent garden spaces.

The Singapore Garden Festival is currently the only international garden exhibition held in tropical regions in the world. It was first held by the National Parks Board of Singapore at the Suntec International Convention and Exhibition Center in December 2006, and held every two years thereafter. In 2018, the 7th Garden Festival was held at Singapore Marina Bay Garden. With an area of up to 2.2 hm^2, it invited 33 well-known garden designers at home and abroad and attracted many private gardening groups and the public. For the first time, the Singapore Garden Festival walked out of the indoor exhibition halls by adding an outdoor exhibition area, so that the public had an opportunity to enjoy a larger and more spectacular display of tropical gardening and floriculture showcase. In addition to these garden festivals, there are also other different garden festivals all over the world, for example, the Westonbirt Garden Show in the United Kingdom and the Ponte de Lima International Garden Festival in Portugal.

图片来源：http://www.dylandscape.com/

花园建造节是花园展中非常特殊的一类，强调的是建造者自己动手建造的过程。建造环节在建筑与园林的教学实践中非常普遍，但是多以小型手工模型为主，大体量的建筑类建造节或建造活动已经在少数院校开展，多以纸板等作为建造材料，在北林国际花园建造节之前，国内尚未出现过花园类的建造节。

建造者在建造节中使用材料来营造空间，将图纸呈现为实物。在这个过程中，建造者可以充分认识材料的性能、建造方式，体验完成的设计、建造的空间环境，基本把握花园的使用功能、人体尺度、空间形态等，感知图纸与实体空间的差异、建造过程对形态的影响、设计与使用功能之间的关系等。

相较于传统意义的花园节，花园建造节更强调的是建造者搭建的步骤和花园逐步呈现的过程。

The garden construction festival is a very special kind of garden exhibition, because it emphasizes the process of construction by builders themselves. The construction process itself is very common in the teaching practice of architecture and landscape architecture, but it is mostly based on small manual models. Large-scale architectural construction festivals or construction activities have been carried out in a few colleges and universities, and most of them use cardboard as construction materials. Before the Beilin International Garden Construction Festival, there was no garden construction festival in China like this.

In such garden construction festivals, the builders use materials to create spaces and convert the drawings into real objects. During the construction process, they can fully understand the properties and construction methods of various materials, experience the completed design and space environment of construction, basically know the gardens' use functions, human scales and spatial forms and perceive the difference between drawings and physical space, the influence of the construction process on form and the relationship between design and use functions, etc.

Compared with the traditional garden festival, the garden construction festival emphasizes the steps of the construction and the gradual presentation of the garden.

北林国际花园建造节要求参加者以原竹和花卉作为核心材料，设计并建造一座小花园。旨在通过材料建构和艺术表达相互交叠运用的建造体验，激发园林学子的创作热情，提高动手能力，弘扬工匠精神，为风景园林行业培养实践创新人才。

建造内容包括一个以原竹为主要材料的景观构筑物，以及与之配套的花园空间。其中，竹构要求以低技术方式进行搭建，占地面积不超过 $8m^2$，限高 4m，应可进入，功能不限，形式随意。竹构不得建设基础，因此构筑物只需凭借地上部分的结构承重。

设计者需充分尊重原竹自然特性和施工技艺特点，探索其作为结构材料、围护材料和装饰材料的多种可能性，寻找最适宜的呈现方式，并结合花卉表达出设计师对花园内涵的理解。

The BFU (Beijing Forestry University) International Garden-Making Festival requires participants to design and build a small garden using raw bamboo and flowers as the core materials. It aims to promote a craftsmanship spirit and cultivate students' creativity through drawings, models and practices and cultivate practical and innovative talents for the landscape architecture subject.

The construction includes a landscape structure with raw bamboo as the main material, and a garden space to match it. The bamboo structure is required to be constructed in a low-tech way, with an area of no more than 8 m^2 and a height limit of 4 m. The garden should be accessible and not limited in function and form. The bamboo structures cannot build foundations, so the structures can only use the above-ground part of the structure to bear the weight.

Designers need to fully respect the natural characteristics of raw bamboo and its construction techniques, explore its multiple possibilities as structural materials, enclosure materials and decorative materials, find the most suitable way of presentation, and combine flowers to express the designer's understanding of garden connotation.

组委会对花园的材料、形式和建造方式进行了充分的限定，不得使用竹子和花卉之外的材料，不得设置水面，不鼓励使用高技能的搭接方式，花园部分要求植物材料仅限花卉和地被植物。希望通过以上限定，增加设计与建造的聚焦性和探索性。

每年从 1 月开始征集竞赛，到 11 月开放活动结束，历时 11 个月左右。确定方案后，参赛团队需在 3 天半的时间内于现场完成 16m^2 微型竹构花园的实体搭建，随后进行为期 1 个月的展览与开放活动。

The organizing committee has fully limited the materials, forms and construction methods of the garden. Materials other than bamboo and flowers are not allowed, water surfaces are not allowed either. High-skilled lap methods are not encouraged. The garden requires only flowers and ground cover plants as plant materials. It hopes that through the above limitation, the focus and the exploratory nature of design and construction will be increased.

Every year, the competition starts calling for entries in January and all activities ends in November. The total duration is about 11 months. After the selected proposals were confirmed, the competition teams need to complete the on-site construction of a 16 m^2 miniature bamboo garden within three and a half days, and then the gardens would carry out exhibitions and opening activities for a month.

从图纸、模型到实物，每个环节都是大学生们实际操练，通过理论与实际结合、材料建构和艺术表达相互交叠运用构筑为园，工匠精神，在此欣飨。这是北京林业大学推进世界一流学科建设，为风景园林行业培养实践创新人才的重要举措，也是北京林业大学师生弘扬工匠精神，为行业发展提供新的灵感和思路的重要探索。

From drawings, models to on-site construction, every link is an indispensable practice opportunity for college students. Through the combination of theory and practice, material construction and artistic expression, the construction of gardens presents the craftsmanship spirit. This is an important measure for BFU to promote the construction of world-class disciplines and to cultivate practical and innovative talents in the landscape architecture industry. It is also a significant exploration for BFU teachers and students to promote the spirit of craftsmanship and provide new inspiration and ideas for the development of the industry.

中心花园——云在亭
CENTRAL GARDEN — SWIRLING CLOUD

大风起兮云飞扬，云在亭取风起云扬之意，占地120m^2，用于本届花园节的信息发布，活动结束后作为师生日常休闲、聚会的户外场所。基地面向校园道路被小树林、灌木、纪念石刻、休闲座椅环绕形成中心广场。

设计从竹材特性入手，发挥原竹柔韧耐弯的优势，采用自由轻松的曲面形态，与校园环境和花园节的氛围相契合。竹亭的各处曲线形开口充分考虑了主要人流以及广场中树木、绿篱和纪念石的位置，方便人们从各个方向自由穿越。竹亭轻盈地跨在原有的绿篱之上，顶部升高束起设为圆形采光口，使建筑与风、阳光和绿植等自然要素内外交融，形成云卷飞扬的意象。

"A gale has risen and is sweeping the clouds across the sky." Inspired by this famous verse in an ancient Chinese poem, the design team has presented a "pavilion of swirling clouds" for the garden festival of BFU. Covering an area of 120 square meters, the pavilion will serve as a hub for information during the festival, and turn into a place for recreation and gathering after the event. Bordering a campus road, the site of the pavilion is a central square surrounded by woods, bushes, memorial stone carvings and benches.

The design approach has taken full advantage of bamboo's features, especially its malleability and bending resistance. Accordingly, a curved form was created for the pavilion, fitting into the campus's context and the atmosphere of the garden festival.The positions of curved openings on the pavilion were determined by the main flow of people and the distribution of trees, bushes and stones on the site so that people can easily walk across the square through the pavilion. The pavilion stretches over the bushes in a light and graceful way. The top of the pavilion narrowed into a round skylight, merging the building with wind, sunlight and green plants, while presenting an image of swirling clouds.

清华大学建筑学院、SUP 素朴建筑工作室
主持建筑师：宋晔皓
项目团队：陈晓娟、孙菁芬、解丹、褚英男、刘梦嘉、于昊惟、师劭航
照明设计：清华大学建筑学院、张昕老师工作室
竹构厂家：安吉竹境竹业科技有限公司
School of Architecture Tsinghua University, SUP Atelier
Principal Architect: Song Yehao
Design Team: Chen Xiaojuan, Sun Jingfen, Xie Dan, Chu Yingnan, Liu Mengjia, Yu Haowei, Shi shaohang
Lighting Design: Tsinghua University X Studio, School of Architecture
Bamboo Contractor: Anji Zhujing Bamboo Technology Co., Ltd.

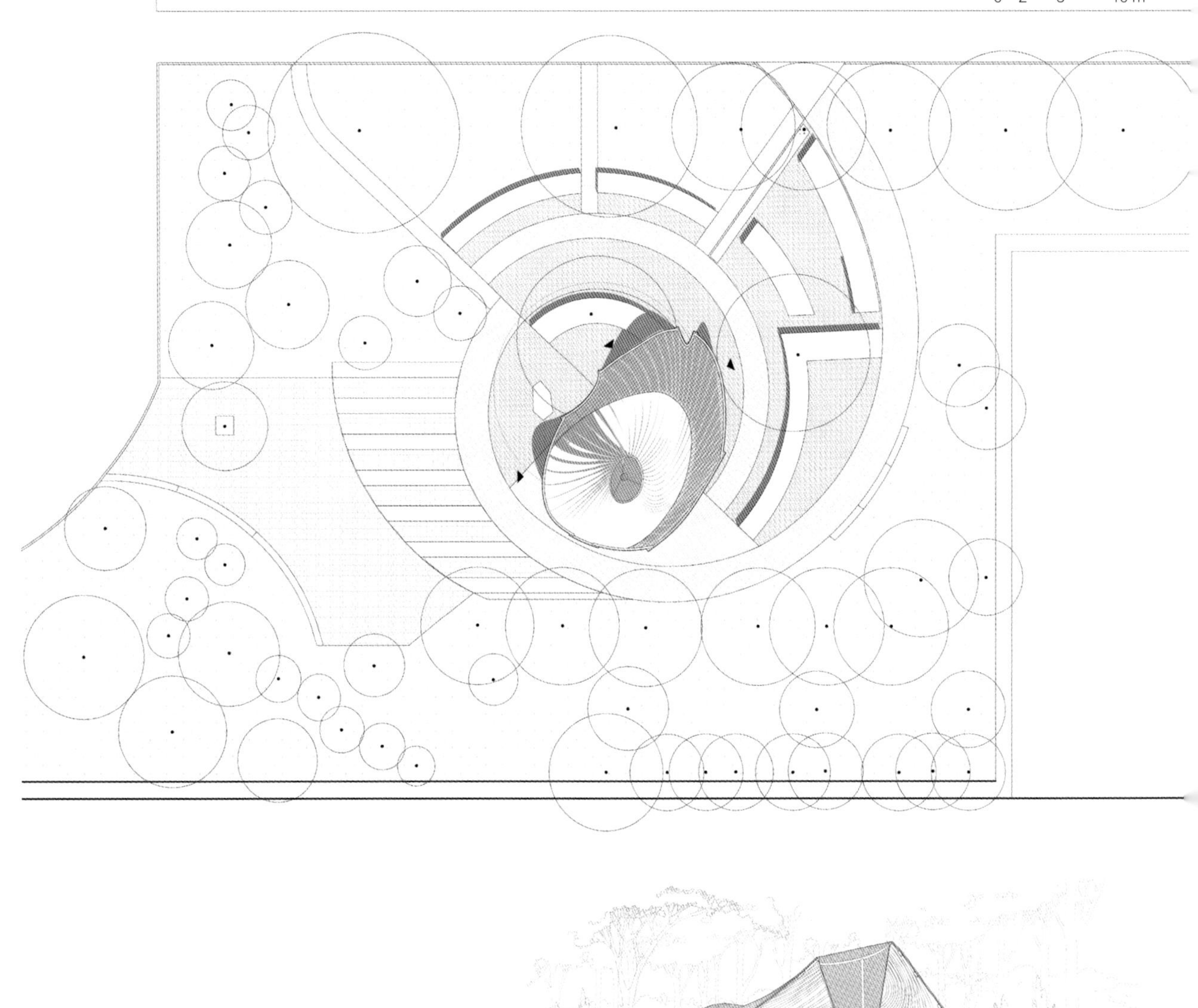
0 2 5 10 m

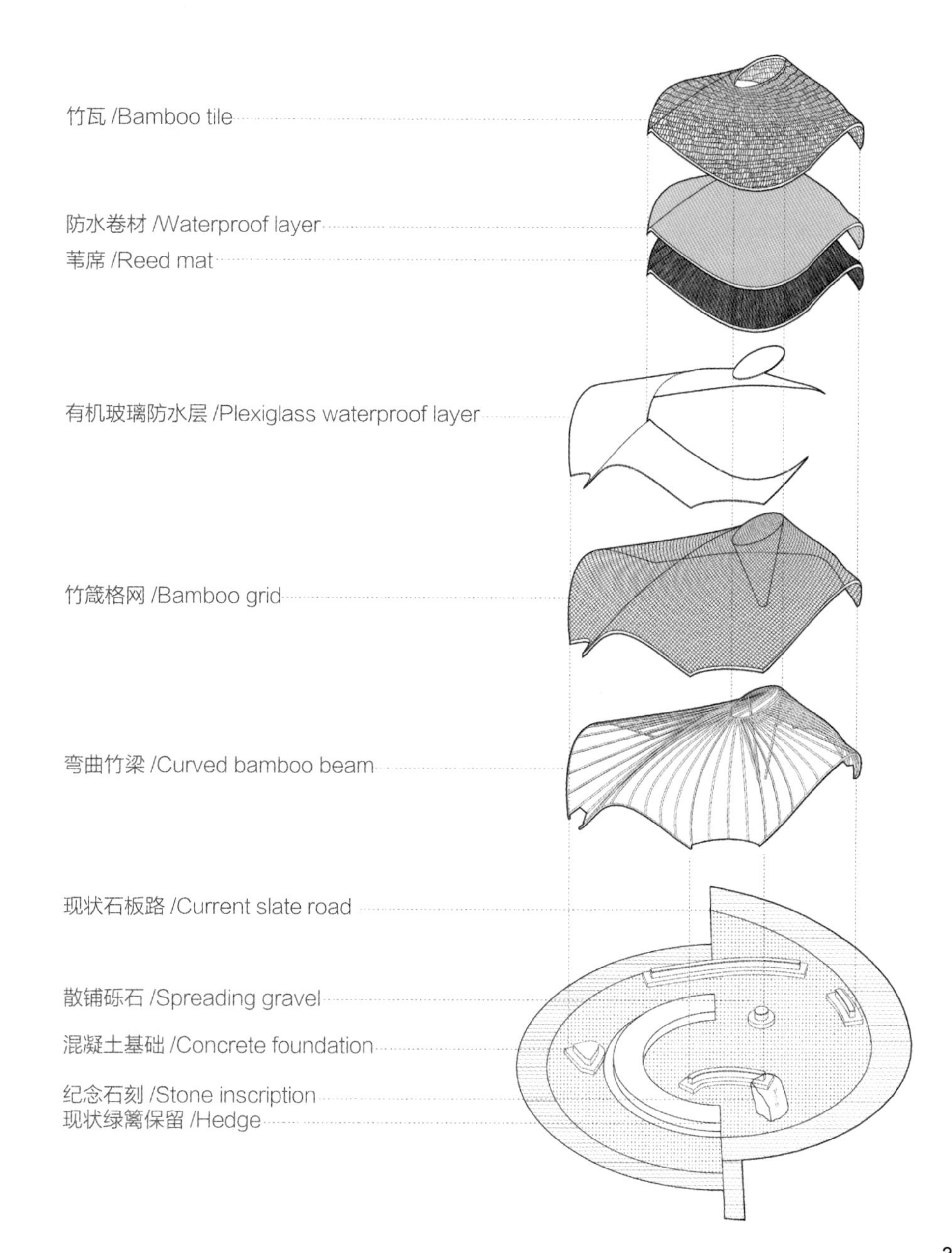
竹瓦 /Bamboo tile
防水卷材 /Waterproof layer
苇席 /Reed mat
有机玻璃防水层 /Plexiglass waterproof layer
竹篾格网 /Bamboo grid
弯曲竹梁 /Curved bamboo beam
现状石板路 /Current slate road
散铺砾石 /Spreading gravel
混凝土基础 /Concrete foundation
纪念石刻 /Stone inscription
现状绿篱保留 /Hedge

第一届（2018）
北林国际花园建造节
The 1st (2018) BFU International Garden–Making Festival

主题：
竹境 · 花园

方案征集时间：
2018 年 3-9 月
实地建造时间：
2018 年 9 月 19-23 日
开放活动时间：
2018 年 9 月 23 日 -10 月 7 日

报名选手：
353 组设计团队，来自 112 所高校，共 1677 人
实地建造团队：
入围团队 8 个、受邀参加团队 7 个

主办单位：
北京林业大学、北京世界园艺博览会事务协调局、中国风景园林学会教育工作委员会
承办单位：
北京林业大学园林学院、安吉竹境竹业科技有限公司、《风景园林》杂志社
支持单位：
国际竹藤组织、北京市花木有限公司、东方园林

Theme:
Landscape of Bamboo

Call for Entry:
March–September, 2018
On-site Construction:
September 19–23, 2018
Opening:
September 23–October 7, 2018

Registered Teams:
353 teams, from 112 universities, a total of 1,677 students
Construction Teams:
8 shortlisted teams and 7 invited teams

Organizers:
Beijing Forestry University, Beijing International Horticultural Exposition Affairs Coordination Bureau, Chinese Society of Landscape Architecture Education Committee
Executive Organizers:
School of Landscape Architecture of Beijing Forestry University, Anji Zhujing Bamboo Industry Technology Co., Ltd, Landscape Architecture Journal
Support Organizers:
International Bamboo and Rattan Organization, Beijing Flowerscape Co., Ltd., Orient Landscape Co., Ltd.

游无穷
TRAVEL IN THE INFINITY

游：

空间的游历——“天”与“地”之间，既有开放畅游之乐，亦有曲径通幽之趣，予人近似山水画的“畅神”体验。每个“山峰”单元间都设有入口，给大人和孩子提供丰富的选择。

无穷：

围绕竹材韧性，将元素串联成一个联动的有机整体。利用座椅“误导”人们坐进“草丛”，使人们亲近花卉，同时带动“山峰”游动，外观与内部空间有“无穷”变化。

Travel:

Space cruise—different entrances between the "mount" units offer diverse "tour lines" for adults and children. Spiritual journey—a fun stroll along the winding path brings people an immersive experience of excursion in the mountains, as if traveling through Chinese landscape paintings, giving people a sense of refreshment.

Infinity:

Seats are used to "mislead" people to sit in the "grass" so as to bring people closer to flowers. Meanwhile, the design connects the elements into an organic whole, which leads the "mountain" units to flow, making the appearance and the internal space infinitely intertwined. Utilize the structure of bamboo tube to extend the design of sliding track. Use bamboo's elasticity to chain all elements into a linkage. The design respects the characteristics of bamboo, and people will enjoy close interaction with bamboo structures.

参赛者学校：南京林业大学
指导老师：程云杉、芦建国
参赛者：赵兮、卢雯、刘静婷、王丽鹏、汪静、张奕、秦琦、费健程
School: Nanjing Forestry University
Advisers: Cheng Yunshan, Lu Jianguo
Participants: Zhao Xi, Lu Wen, Liu Jingting, Wang Lipeng, Wang Jing, Zhang Yi, Qin Qi, Fei Jiancheng

Travel | in the Infinite

感觉身体被掏空
FEELING HOLLOWED OUT

美国 The LAB 于 2017 年发布《各专业平均睡眠时间榜单》，建筑学以平均每天睡眠 5.28 小时位居首位，赶超法学 6.29 小时、医学 6.26 小时。在繁重的学业压力下，“三天一小熬，五天一大熬”成为建筑学学生的日常，睡眠惺忪、蓬头垢面更是建筑专业学生熬夜画图时的常态。或追求完美一直奋战到汇报前一刻的紧张，或偷个小懒把图交上后回宿舍睡大觉的侥幸，在这个构筑物里都可以找到自己和他们的影子。

According to the report "How many hours' sleep dose your major gets?" released by the LAB in 2017, architects ranks first with the minimal sleep of about 5.28 hours' sleep per day. Staying up every three or five days has become a common pattern of architecture students. No matter what type of landscape architect you are, you can find your own shadow here.

参赛者学校：清华大学
指导老师：朱育帆、吕回
参赛者：王劭仪、许清如、高泽宁、陈度、李熙盈、李雨萌、王心语、林晨涛
School: Tsinghua University
Advisers: Zhu Yufan, Lv Hui
Participants: Wang Shaoyi, Xu Qingru, Gao Zening, Chen Du, Li Xiying, Li Yumeng, Wang Xinyu, Lin Chentao

建筑生日常

Everyday life of an architecture student

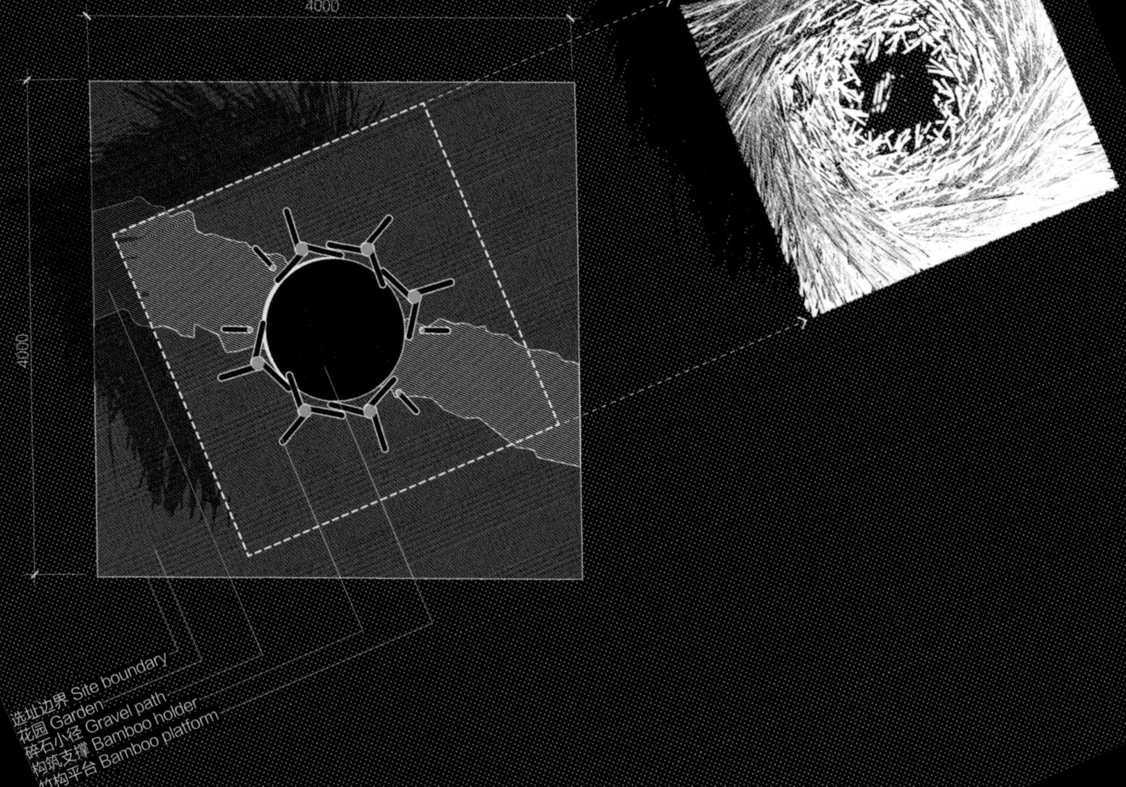
4000
4000
选址边界 Site boundary
花园 Garden
碎石小径 Gravel path
构筑支撑 Bamboo holder
竹构平台 Bamboo platform

花鸟卷
A HANDSCROLL OF FLOWER-AND-BIRD

本方案设计理念来自中国传统文化与当代校园文化的融合，设计形态来源于书卷和花鸟。

整体设计分为三个部分：
第一，竹编组成的结构犹如烙印花鸟的画卷，体现“画境”的含义；
第二，小品也是给人提供休憩和交流的座椅，使人犹如归鸟回到这里，体现“归鸟”；
第三，整体花境与竹编结构相结合体现出“花鸟相闻”的和谐环境。

The design expresses the integration of Chinese traditional culture and modern campus life. The inspiration comes from the Flower and Bird scroll.

The overall design has three characteristics:
First, the bamboo structure is like a scroll printed with flowers and birds and reflects the "picturesque scene".
Second, the structure is equipped with seats which provide pleasant space for people to rest and chat so that it could make people linger and reflect the concept of "birds returning".
Third, the combination of flower border and bamboo structure expresses the harmonious atmosphere of birds and flowers.

参赛者学校：北京林业大学
指导老师：冯潇、郑小东、段威
参赛者：周超、王宏达、李爽、赵可极、王亚迪、税嘉陵、蔡怡婷、何启之
School: Beijing Forestry University
Advisers: Feng Xiao, Zheng Xiaodong, Duan Wei
Participants: Zhou Chao, Wang Hongda, Li Shuang, Zhao Keji, Wang Yadi, Shui Jialing, Cai Yiting, He Qizhi

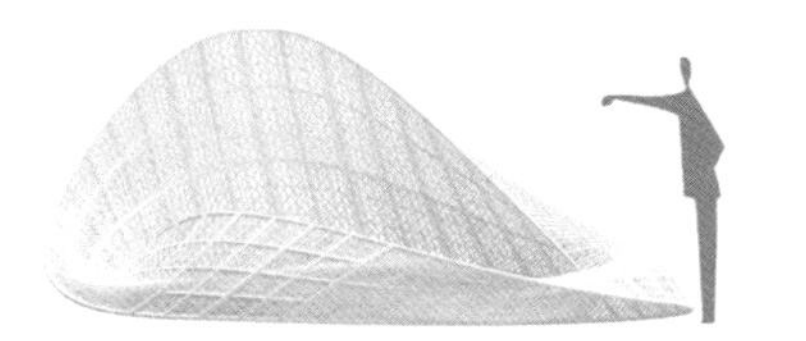

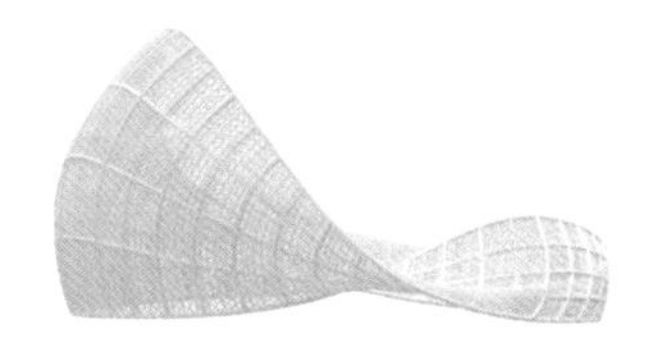

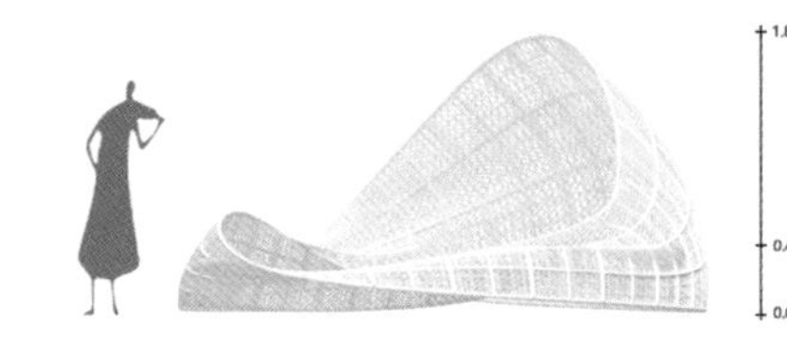
1.80m
0.45m
0.00m

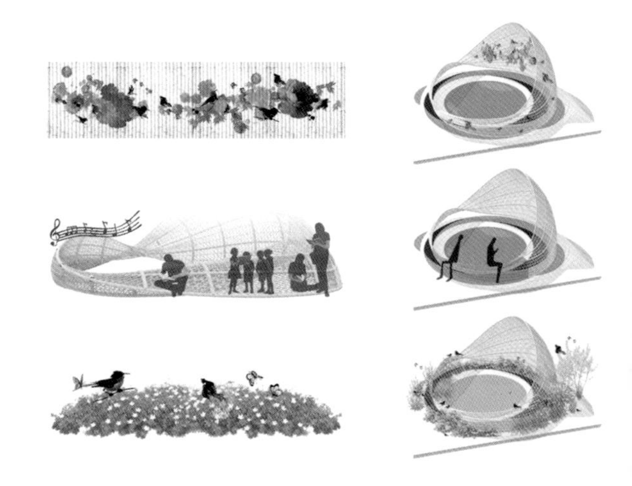

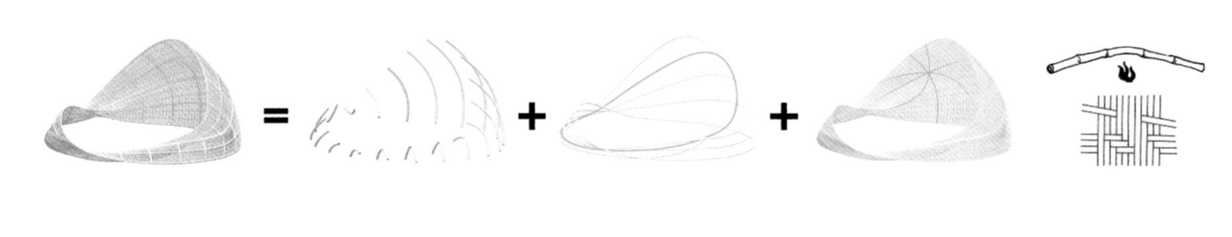

0.5
1.5
0
1
2m
设计场地

方秩律
THE CUBE

方非方，竹非竹。
四面疏影藏已匿，九柱列阵古与今。
君欲探内乾坤事，芒草萋萋风弄伊。

方秩律源于弗兰姆普敦现代主义建构理念，延续海杜克空间立方体构成原则，探寻六面一体的竹构空间营建语言，重塑竹境使命，实现建构逻辑、结构逻辑、空间逻辑与人文情怀的理性融合。传统工法加以现代演绎，本土材料映射时代内涵，栽植空间延续自然属性，内外对望暗含伦理秩序，节奏脉动凸显场所张力。

Cubic, it is not square; Bamboo made, it is not bamboo. The shadow is hidden from all sides. The nine columns present ancient and modern. He wants to find out what's going on inside. He is kidded by grass in the wind.

The cube, inspired by Frampton's modern construction theory and Hejduk's cube constitution principle, is designed to clarify the construction language of space inside the bamboo structure. Designers endeavor to realize the combination of construction logic, structure logic, space logic and human concerns. Traditional methods are performed in a modern context. The adoption of local materials reflects a new time connotation. The natural quality of bamboo extends from an intimate interior to an urban exterior. Orders are manifest in the in-and-out communication, as the site is tensioned by the rhythm of bamboo components.

参赛者学校：东南大学
指导老师：李哲
参赛者：黎颖琳、曹息、常晓旭、翟志雯、吴宇坤、李鑫、谢祺铮
School: Southeast University
Advisers: Li Zhe
Participants: Li Yinglin, Cao Xi, Chang Xiaoxu, Zhai Zhiwen, Wu Yukun, Li Xin, Xie Qizheng

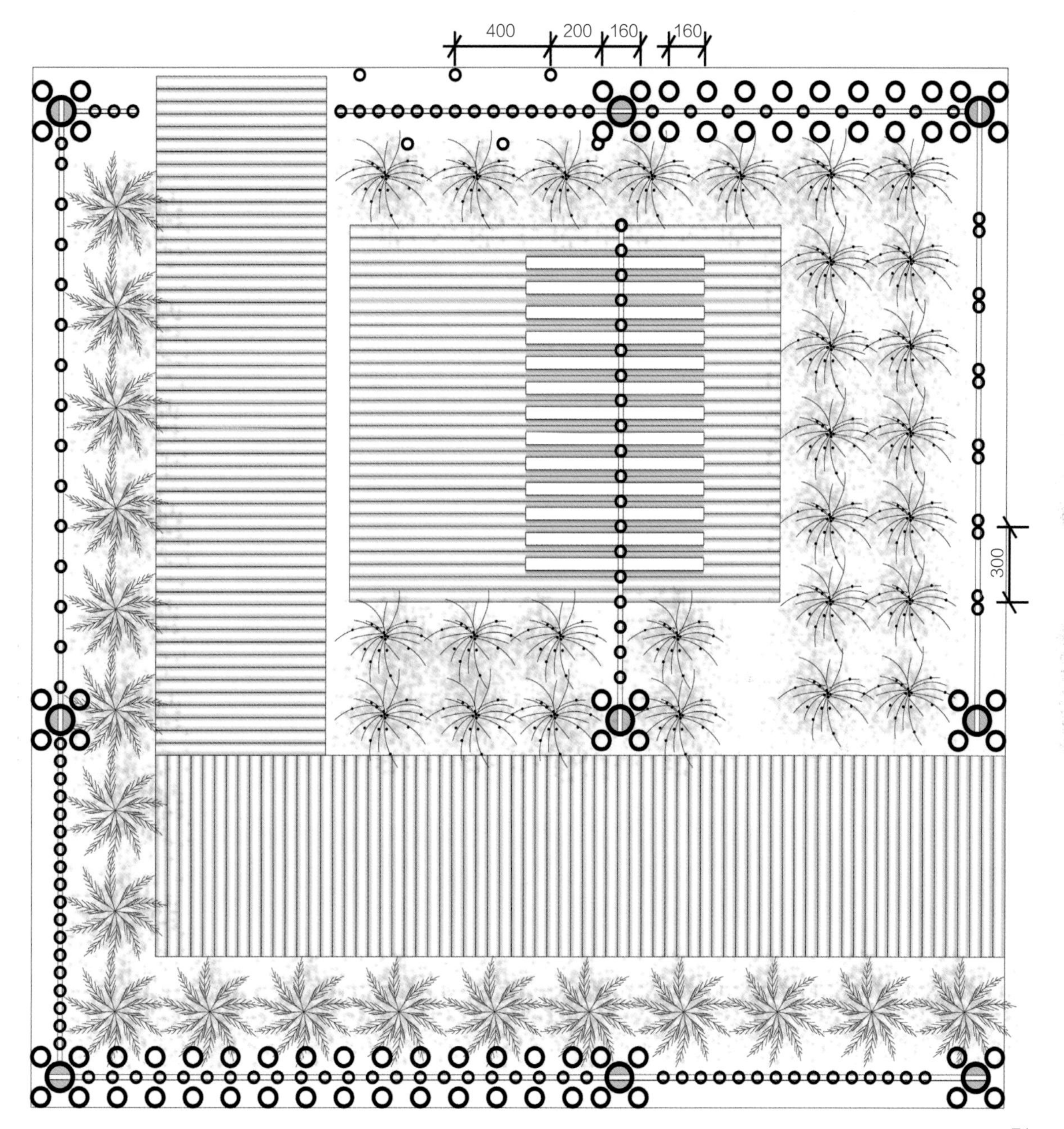
400
200
160
160
300

罄·篁
ECHOING ROAM

竹子有挺直、柔韧、中空、成节的特性，立于地而自弯，迎风相击而鸣清音，横剖于节上则可盛物，成簇成林则遮清荫，其形式变换无穷，带给人丰富的多感官体验。

在纯粹的正交体系下，竹与花的万般变化含于内，又透过层层竹隙若隐若现，像是一座竹与花生长出的迷宫。花园中可以栽竹，竹中可以生花，此是为“竹境花园”。

Bamboo has the characteristics of straightness, flexibility, hollowness and jointness. Bamboo naturally bends due to its flexibility, striking each other in the wind and makes a melodious sound. Cross-cutted above the knot, it can hold objects. By Clustering together they can form a grove, which produces a refreshing shade. An infinite variety of shapes bring people a rich and multi-sensory experience.

In the purely orthogonal structure, the variations of bamboo and flowers are hidden within, and the layers of bamboo looms make it like a labyrinth of bamboo and flowers. Bamboo can be planted in the garden, and within bamboo flowers can bloom, which makes a real "Bamboo Garden".

参赛者学校：同济大学
指导老师：周宏俊、许晓青、沈洁、汪洁琼
参赛者：陈昱萌、陈茜、刘锟山、崔迪、李晓薇、鄔沅蓉
School: Tongji University
Advisers: Zhou Hongjun, Xu Xiaoqing, Shen Jie, Wang Jieqiong
Participants: Chen Yumeng, Chen Xi, Liu Kunshan, Cui Di, Li Xiaowei, Wu Yuanrong

罄·篁

举头望明月
LOOK UPON THE MOON

《竹取物语》的传说，让日本文化中的竹与月成为两个相关联的意象。正值中秋时节，本方案以“竹中月”为概念，运用日本庭师的打结技巧进行构件间的连接，围合出圆柱形的空间。方案的设计考虑到人赏月的视线角度，空间整体呈斜向上开口。圆内设有围栏，其高度可供肘部自然倚靠。圆中摆放芒草，象征所在的大地，与空中明月遥相呼应。

"The tale of the bamboo cutter" gives a relationship between bamboo and moon in Japaness culture. This scheme chose "Moon in Bamboo" as the concept as the Mid-Autumn Festival is at the time, and the traditional Japanese Gardeners' knotting skills are used to connect the components to form a cylindrical space. Considering a comfortable viewing angle for moon watching is required, the whole space is inclined upward. The fence inside the bamboo circle allows the elbow to lean naturally. Miscanthus fills up the remaining space, symbolizing the earth we stand on.

参赛者学校：日本国立千叶大学
指导老师：霜田亮祐
参赛者：宇都宫青流、胡博文、村上善明、佐佐木圭殿、酒井孝浩、田木日奈子、向吉真央、石渠
School: Chiba University
Advisers: Ryosuke Shimoda
Participants: Kiyoharu Utsunomiya, Hu Bowen, Yoshiaki Murakami, Kei Sasaki, Takahiro Saka, Hinako Tagi, Mao Mukoyoshi, Shi Qu

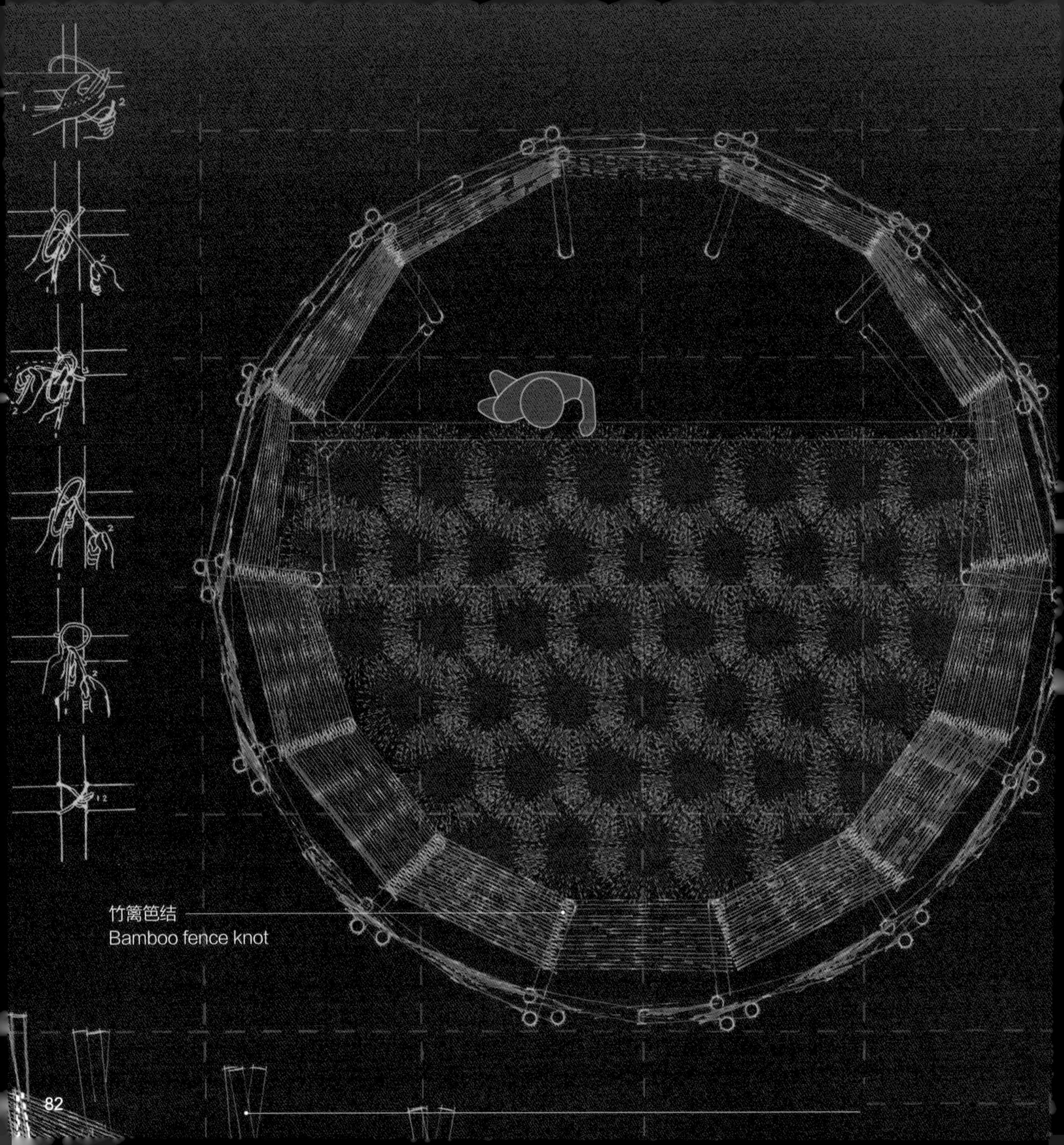
竹篱笆结
Bamboo fence knot

三脚架
Tripod knot

曲水流山
WINDING RIVER & FLOWING MOUNTAIN

本设计方案名曰“曲水流山”，意在城市中通过竹材创造自然的山水缩影。“曲水”寓意植物与蜿蜒的空间路径形成的流水之感，“流山”寓意竹构的空间体量形成的山体之势，两者相映成趣。方案的灵感来源于“曲水流觞”一词，是修禊的主要活动。王羲之于兰亭修禊，用曲水流觞来赋诗作文，成千古风流范本，影响后世。此次花园搭建节，以兰亭修禊图为切入点，着眼掇山和曲水，用新的形态结构加以转换和再现。

This project is called "Winding River & Flowing Mountain", which is built with bamboo, intending to create a natural landscape microcosm in the city. "Winding River" implies the feeling of flowing water formed by plants and space paths, "Flowing mountain" implies the formation of space volume, providing a pleasant experience to visitors. The inspiration of the scheme comes from the word "Water-Festival Game", which was an important activity called "Xiu qi". Wang Xizhi, who was a great calligrapher, wrote poems and compositions while playing the game. It became a model of the past and influenced the later generations. With the Orchid Pavilion as the source of inspiration, the design took mountains and meandering water as image to transform and reproduce the morphological structure of the work.

参赛者学校：中央美术学院
指导老师：侯晓蕾、钟山风
参赛者：贾思屹、戴湖浩、王若飞、李师成、孙昆仑、吴子旸、李博、骆驿
School: Central Academy of Fine Arts
Advisers: Hou Xiaolei, Zhong Shanfeng
Participants: Jia Siyi, Dai Huhao, Wang Ruofei, Li Shicheng, Sun Kunlun, Wu Ziyang, Li Bo, Luo Yi

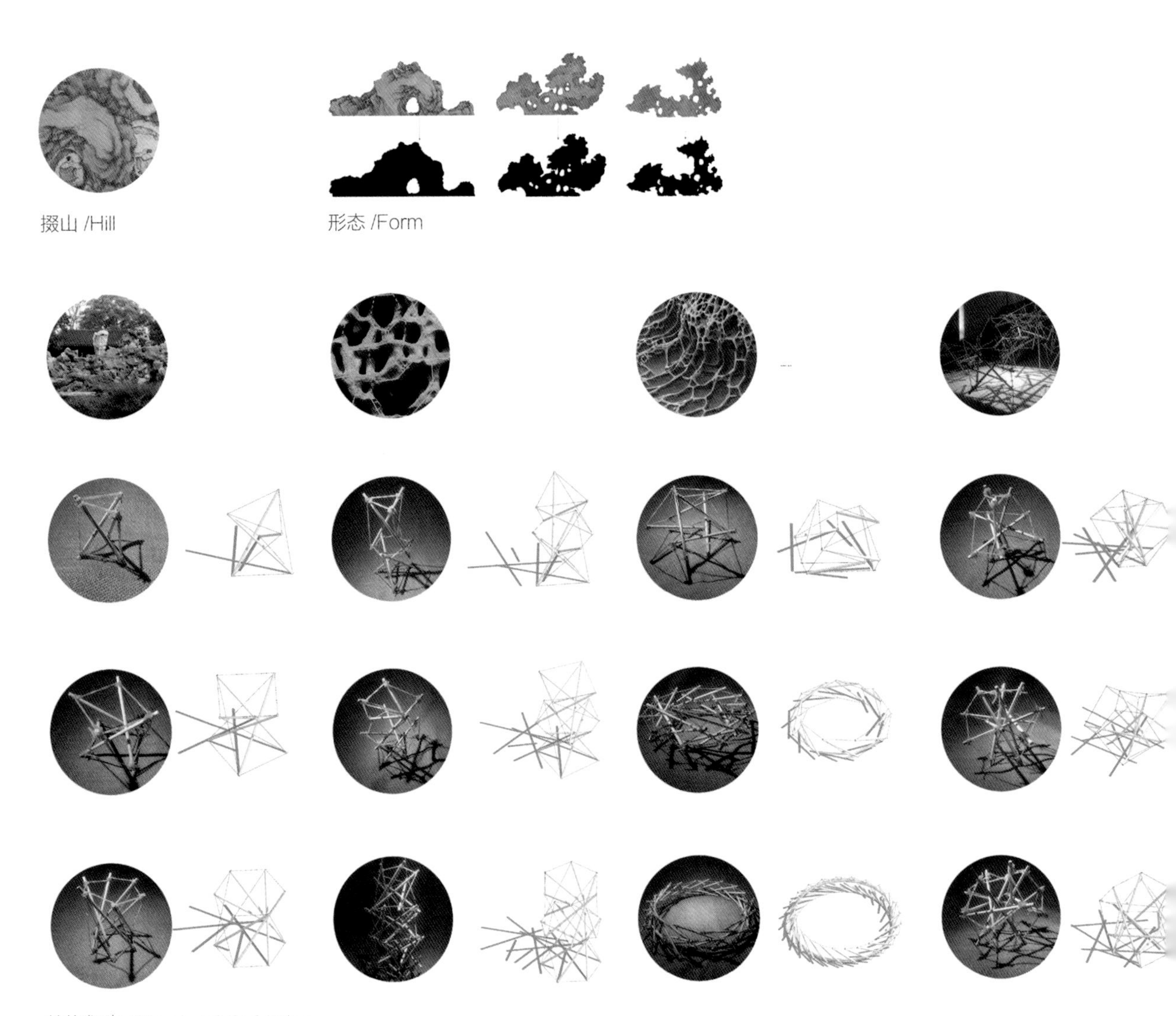

结构衍生 /Structural derivation

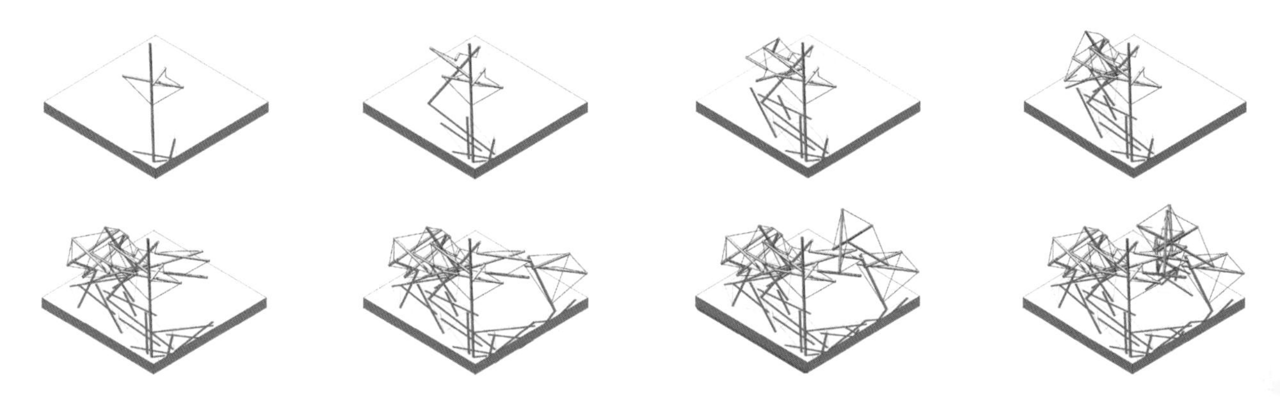

形体生成 /Form formation

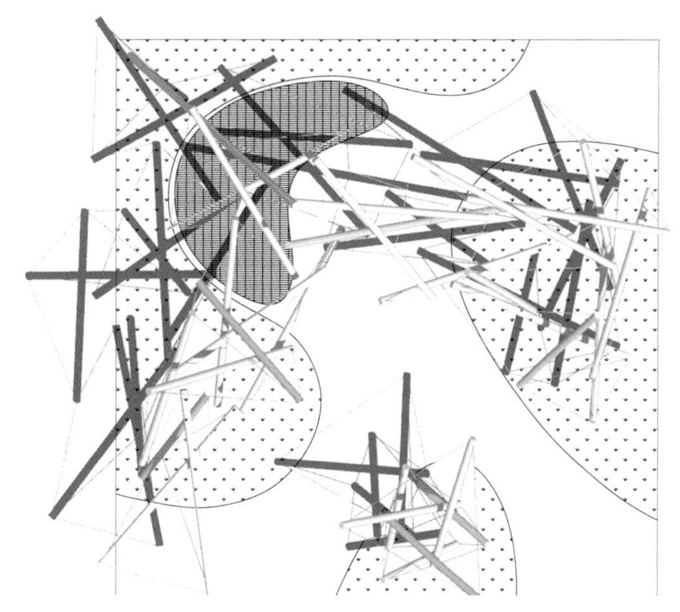

格列佛转角花园
GULLIVER'S CORNER GARDEN

我们理想中的花园能够像格利佛魔法隧道一样趣味无穷，来者能够抛开周身的琐事，以孩子般纯真的眼光观察世界。在逐渐扩大的空间里期待愈加广阔的天空；在逐渐缩小的空间里好奇越发细微的精彩。设计此花园时，我们利用等比例缩小的体块旋转叠加，营造出大小变换的空间感受；通过阶梯式的有序排列，搭配种类丰富的盆栽花卉，创造丰富多彩的视觉体验。

We design an ideal garden where visitors can put trifles aside and bring childlike innocence back for a while, like the Gulliver's magic tunnel. The garden provides different spatial feelings. By walking through the orderly arrangement of steps and a variety of potted flowers, visitors will enjoy a colorful and fantastic visual experience. Though it's a narrowing garden, people can slow down to experience wonderful life. It's also a growing garden, which keeps people's imagination inspiring.

参赛者学校：浙江农林大学
指导老师：包志毅
参赛者：胡真、马超、杨翔、沈天宇、李铜攀、叶晨浩
School: Zhejiang A & F University
Advisers: Bao Zhiyi
Participants: Hu Zhen, Ma Chao, Yang Xiang, Shen Tianyu, Li Tongpan, Ye Chenhao

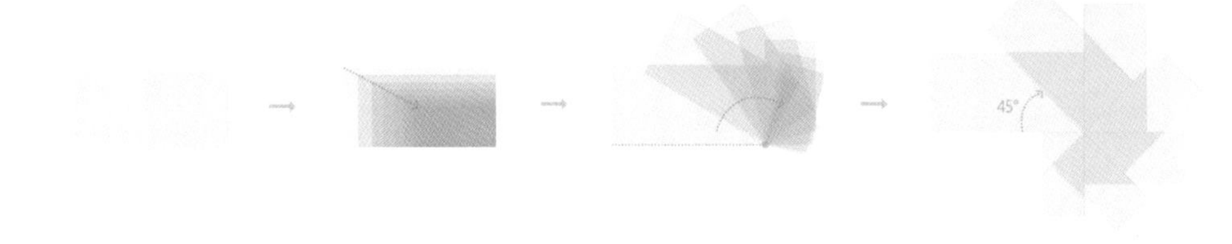
45°

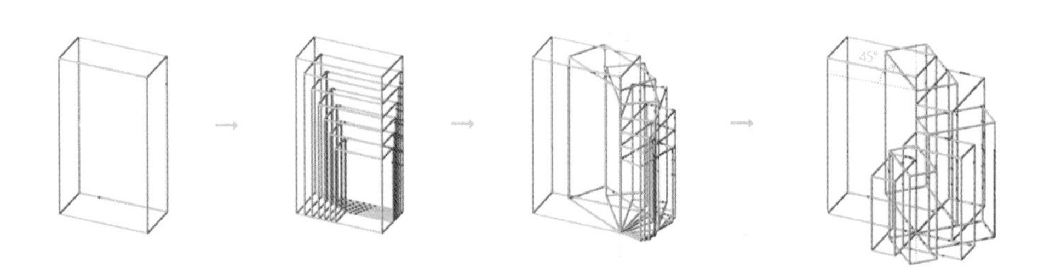
45°

8

竹阁
BAMBOO VIEW FINDER

现时都市人每日都离不开科技产品，却忽略了许多自然美景。因此，在北京林业大学这个充满不同景观的如同大自然的环境中，我们通过提供一个地标性的竹亭，让人们的心灵可以重新回到大自然中，暂时放下手机，欣赏花草树木，静思。

亭子处于一个能让大众轻易看到的位置，因而成为一个集合点。大众的视线都会被位于中心点的石头吸引，亭子的设计使其可以从多个角度被看到。且亭子能够与地点环境在四季的颜色上有较大的对比，以吸引学生视线并引导学生注意园内景色。

Nowadays, city dwellers are inseparable from electronic devices every day, but they are more or less lack of the cognition of natural beauty. Therefore, in the campus environment that is full of different landscapes, this project provides people with a landmark bamboo pavilion that allows them to temporarily put down their mobile phones and return to nature, so that they can meditate and appreciate flowers, trees.

The pavilion is placed in a location that is well visible to the public so that possess the potential to become a meeting point. The stones located at the center will catch the attention of the public; and the design of the pavilion will become a focal point, allowing people to see it from any angle. Besides, the pavilion has a contrast with the surrounding environment. Therefore, the work can draw student's attention to have a look and guide their vision to the beautiful view of the park.

参赛者学校：香港珠海学院
指导老师：夏珩、肖亚娜
参赛者：朱俊賢、趙立明、田俊杰、容紫晴、程琬淇、賴奕燊
School: Chu Hai College of Higher Education, Hongkong
Advisers: Xia Heng, Xiao Yana
Participants: Chu Chun Yin, Jau Li Ming, Tin Chun Kit, Yong Tsz Ching, Ching Yuen Ki, Lai Yik Sun

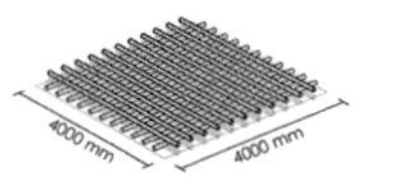
4000 mm
4000 mm

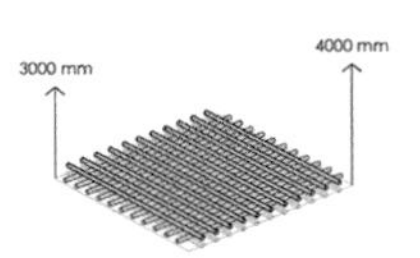
3000 mm
4000 mm

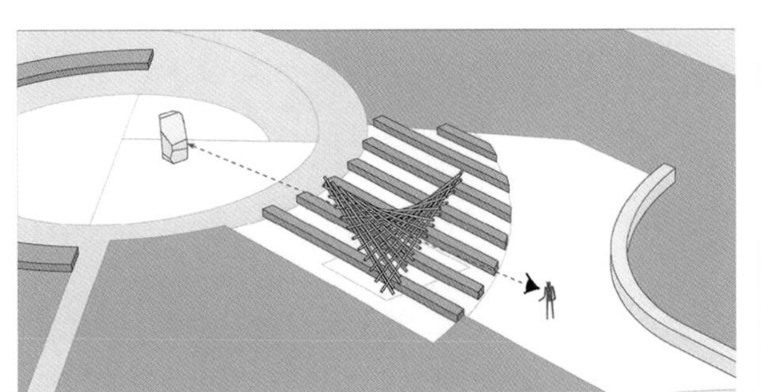

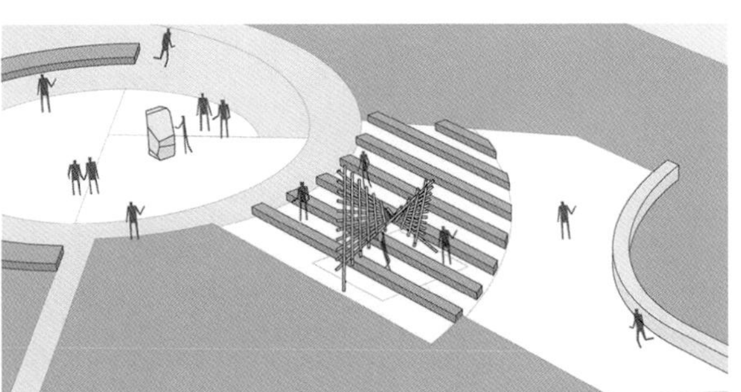

北京林业大学有很多美妙的隐藏景观，在繁忙的学业中，学生很容易忽略这些景观。提供一个地标性的亭子让学生在经过时停下脚步或吸引学生将这个亭子用作等候及集合的地方。同时通过亭子的形状，引导学生发现北京林业大学内的隐藏景观，令学生在欣赏美妙环境的同时得到放松。

BFU obtains plenty of wonderful hidden landscape which students always ignore it in their busy academic studies. Therefore, providing a landmark pavilion that lets the students rest or attracts students to use this pavilion as a gathering point. At the same time, the shape of the pavilion will guide the students to discover the hidden landscapes in the university, so that students can relax while enjoying the wonderful environment.

落竹
FALLING BAMBOO

“落竹”这一作品的顶部由天上掉下来的竹子组成，这给予参观者一个新的视角，并重新发现竹子作为一种材料的特性。作品的顶部在滑轮的帮助下可以从一边移动到另一边，参观者可以通过移动竹子，将自身融入周围环境，并自己创造各种各样的风景。期待这个装置能为参观者提供新的视角、动态景观和独特的体验。花园部分也将作为一种媒介，在视觉上表现新的视角，加强不同寻常的体验。

The ceiling of the "Falling Bamboo" consists of bamboos with the intentional form of “falling down from the sky”. This allows visitors to see a new perspective, and provides a rediscovery of bamboo as a material. The pavilion's ceiling moves from one side to another with the help of pullies. Participants are able to move the bamboos by themselves to create various landscapes. This activity subsidizesa fresh perspective given by previous settings, providing people with interactive landscapes and unique experience. The plantation also serves as a medium to actively show the new perspective, strengthening the unusual experience.

参赛者学校：庆熙大学
指导老师：金振午、閔丙旭
参赛者：權銀娥、張僖庭、申惠琳、朴眞率、金沿禹、李有珍
School: Kyung Hee University
Advisers: KIM, JIN-OH; MIN, BYOUNG WOOK
Participants: Euna Kwon, Hee Jeong Jang, Hye Lim Shin, Jinsol Park, Yeon Woo Kim, Yujin Lee

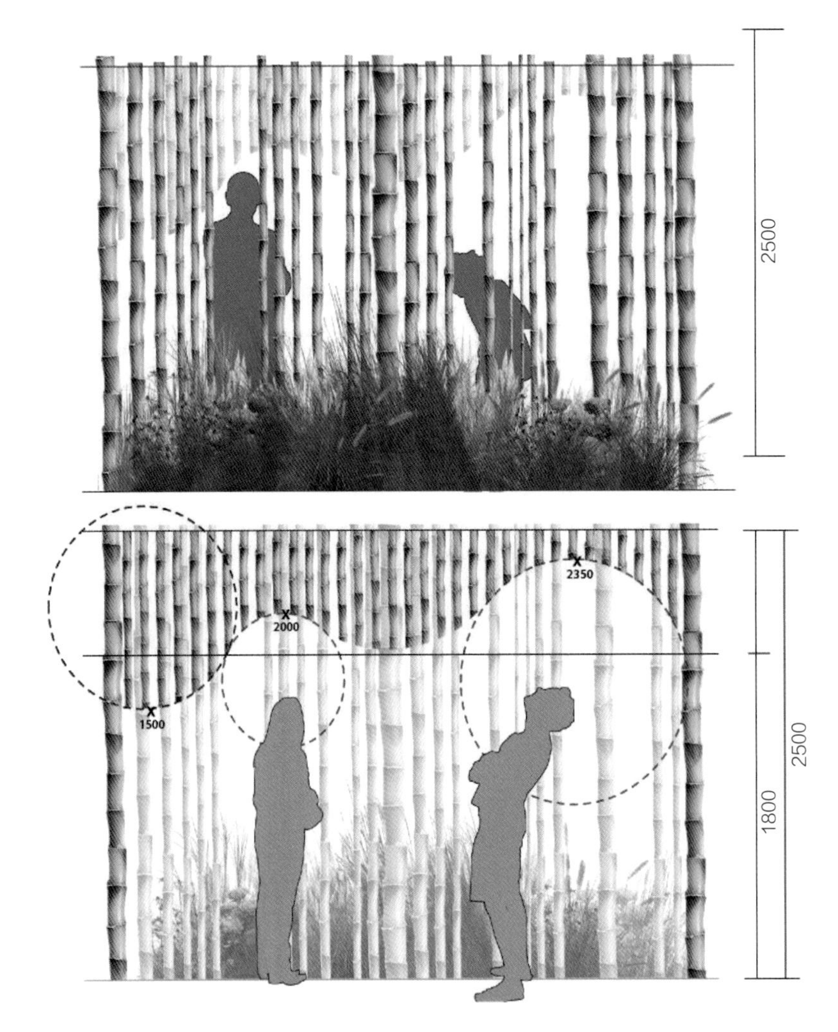
2500
2350
2000
1500
1800
2500

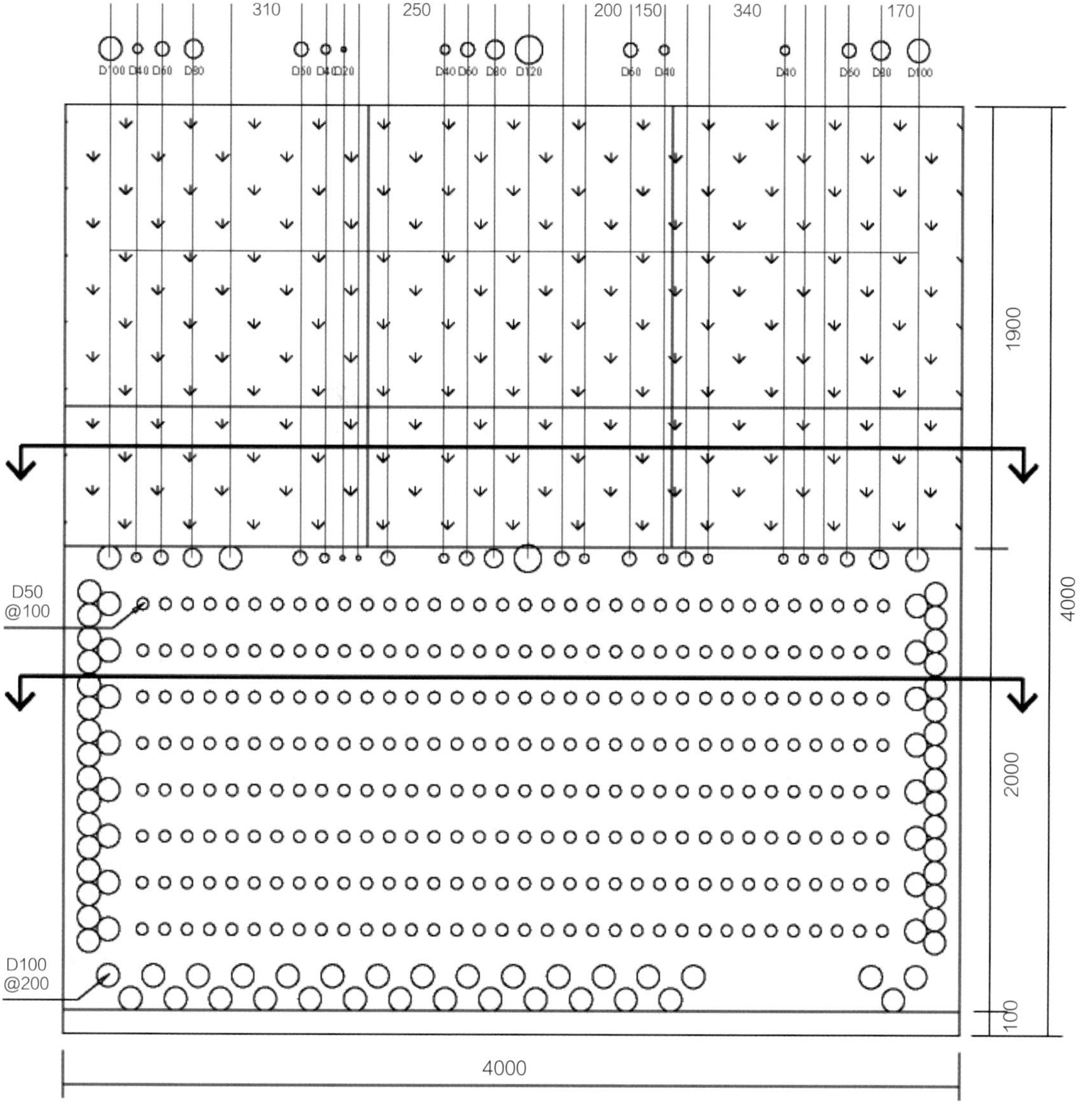

310 250 200 150 340 170

D100 D40 D60 D80 D60 D40 D20 D40 D60 D80 D120 D60 D40 D40 D60 D80 D100

D50
@100

D100
@200

1900
2000
100
4000

4000

织 · 阵
WEAVING IN ARRAY

本方案意在学校环境中创造具有记忆感与庇护感的自然氛围。结构方面，方案采用三组弯竹形成的弯竹支撑体系结合成拱廊，由于开创性地取消了竖向构件，结构在内部与外部产生了自然化的体验，给予人丛林般原始的保护感。景观方面，方案创新性地采用了立体的景观布置——竹构内部花境营造私密的神秘花园。

This scheme wants to create an impressive atmosphere with natural elements. In terms of structure, the scheme uses an arched bamboo support system formed by three groups of bamboos to form an arcade. As a result of the groundbreaking elimination of vertical components, the structure produces a naturalized experience inside and outside, giving the jungle a primitive sense of protection. As for landscape, this scheme has innovatively adopted a three-dimensional landscape arrangement: inside the bamboo, the flowerboard create a private and secretive garden.

参赛者学校：清华大学、北京林业大学
指导老师：朱宁、王美仙
参赛者：孙照人、刘泽洋、宋天意、曾暐翔、李雁晨、刘恋、诸禹圻、黄子盈
School: Tsinghua University, Beijing Forestry University
Advisers: Zhu Ning, Wang Meixian
Participants: Sun Zhaoren, Liu Zeyang, Song Tianyi, Zeng Weixiang, Li Yanchen, Liu Lian, Zhu Yuqi, Huang Ziying

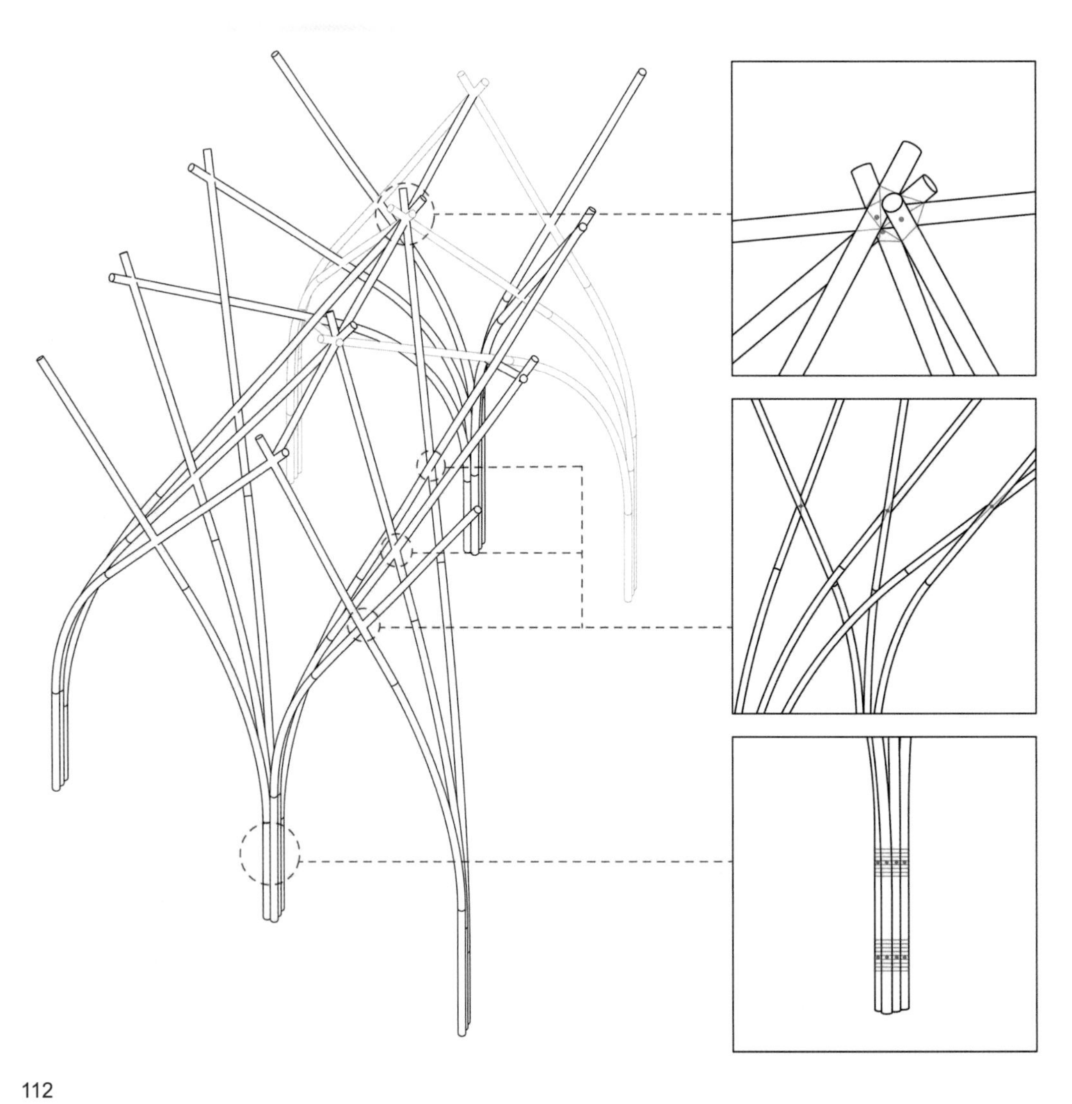

林中漫舞

DANCING IN THE FOREST

草长莺飞的四月，姑娘们乘着暖暖的春意在林间与彩蝶漫舞，散发出青春的朝气与活力。方案的灵感来源于林中漫舞的姑娘，用轻巧柔韧的竹构展现出飞旋的舞裙，花境成为她的舞台。设计最大限度地展现了竹子的柔韧与韵律，并充分利用竹篾的轻巧与灵动，共同构成飘逸的竹构景亭。蜿蜒的花境使竹亭内外空间流动起来，随着光影的漫动，融融的竹境，一片柔美。

In April when the grass warblers fly, the girls dance with the butterflies in the woods in the warm spring, giving off the vigor and vitality of youth. Inspired by the dancing girl in the woods, the scheme uses a light and flexible bamboo structure to display a flying skirt, and the flower border becomes her stage. The design maximizes the flexibility and rhythm of bamboo, and makes full use of the lightness and agility of bamboo strips to form an elegant bamboo pavilion. The winding flower border moves the bamboo pavilion inside and outside the space. The harmonious bamboo environment is soft and beautiful.

参赛者学校：天津大学
指导老师：王洪成、胡一可
参赛者：郭茹、王雪睿、杨宁、曹烨琪、孙雅伟、张颖、陈丽君、李伊侬
School: Tianjin University
Advisers: Wang Hongcheng, Hu Yike
Participants: Guo Ru, Wang Xuerui, Yang Ning, Cao Yeqi, Sun Yawei, Zhang Ying, Chen Lijun, Li Yinong

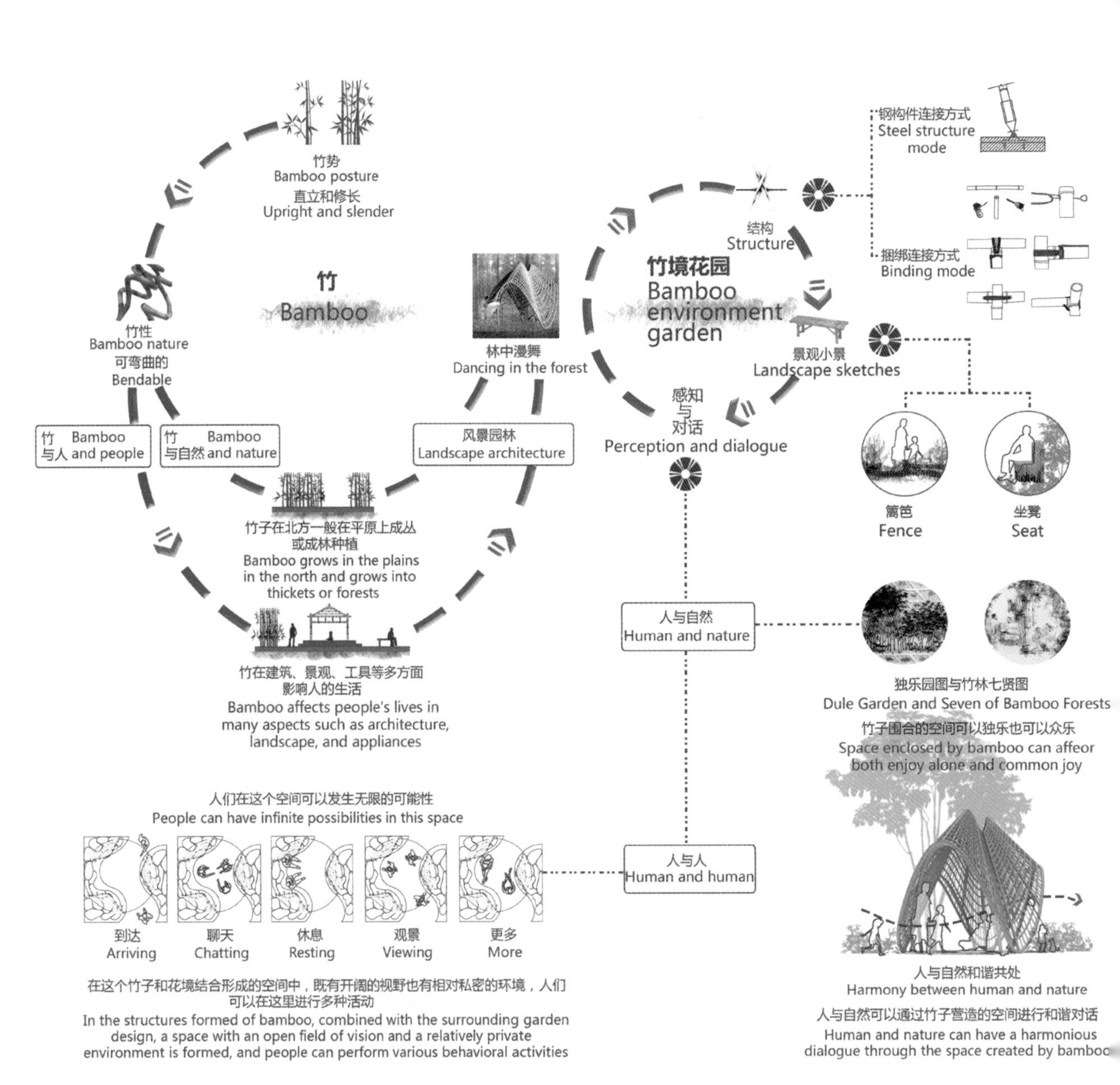

竹势
Bamboo posture
直立和修长
Upright and slender
竹
Bamboo
竹性
Bamboo nature
可弯曲的
Bendable
林中漫舞
Dancing in the forest
竹境花园
Bamboo environment garden
结构
Structure
钢构件连接方式
Steel structure mode
捆绑连接方式
Binding mode
景观小景
Landscape sketches
感知与对话
Perception and dialogue
竹 Bamboo 与人 and people
竹 Bamboo 与自然 and nature
风景园林
Landscape architecture
篱笆
Fence
坐凳
Seat
竹子在北方一般在平原上成丛或成林种植
Bamboo grows in the plains in the north and grows into thickets or forests
人与自然
Human and nature
竹在建筑、景观、工具等多方面影响人的生活
Bamboo affects people's lives in many aspects such as architecture, landscape, and appliances
独乐园图与竹林七贤图
Dule Garden and Seven of Bamboo Forests
竹子围合的空间可以独乐也可以众乐
Space enclosed by bamboo can affeor both enjoy alone and common joy
人们在这个空间可以发生无限的可能性
People can have infinite possibilities in this space
人与人
Human and human
到达
Arriving
聊天
Chatting
休息
Resting
观景
Viewing
更多
More
在这个竹子和花境结合形成的空间中，既有开阔的视野也有相对私密的环境，人们可以在这里进行多种活动
In the structures formed of bamboo, combined with the surrounding garden design, a space with an open field of vision and a relatively private environment is formed, and people can perform various behavioral activities
人与自然和谐共处
Harmony between human and nature
人与自然可以通过竹子营造的空间进行和谐对话
Human and nature can have a harmonious dialogue through the space created by bamboo

3200
4000

开合之间
OPEN AND CLOSE

“开合之间”方案受到西关大屋三道门的启发，利用视线与动作的阻隔定义空间性质。本方案提炼角门、趟陇和木门的生效原理并将其运用到设计中，表现为旋转门与趟栊的组合。

人在行进过程中与门发生互动并不断重新定义空间。

"Open and Close" is inspired by the three-door of Xiguan House in Guangzhou. It defines spatial associations by separating views and movements. This scheme adopts the practice of Horn gate, Tanglong and Wooden gate to guide the design, which performs as the combination of the revolving door and Tanglong.

People can interact with the doors and redefine the space when they are moving.

参赛者学校：华南农业大学
指导老师：江帆影，林毅颖
参赛者：邹嘉铧，吴林倩，金晶，黄冰怡，陈子霖，陈雪，陈漫婷，李欣怡
School: South China Agricultural University
Advisers: Jiang Fanying, Lin Yiying
Participants: Zou Jiahua, Wu Linqian, Jin Jing, Huang Bingyi, Chen Zilin, Chen Xue, Chen Manting, Li Xinyi

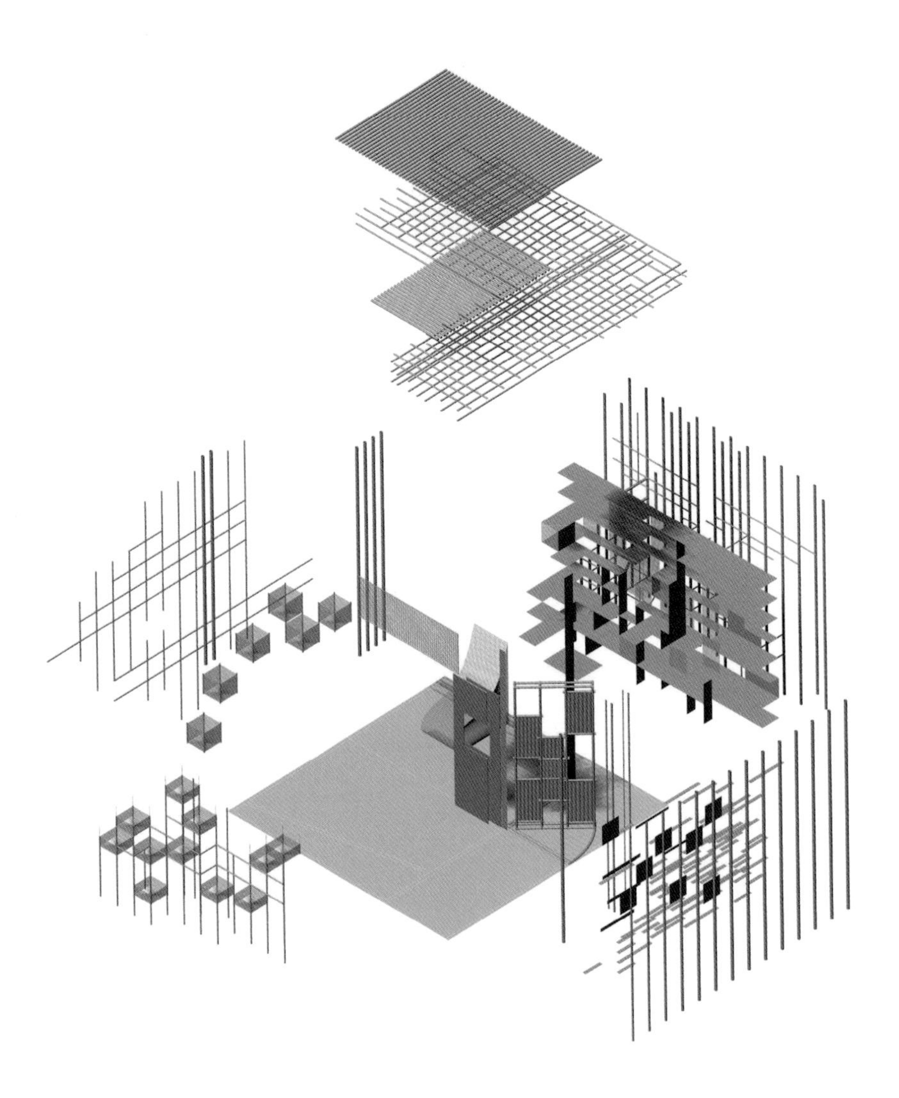

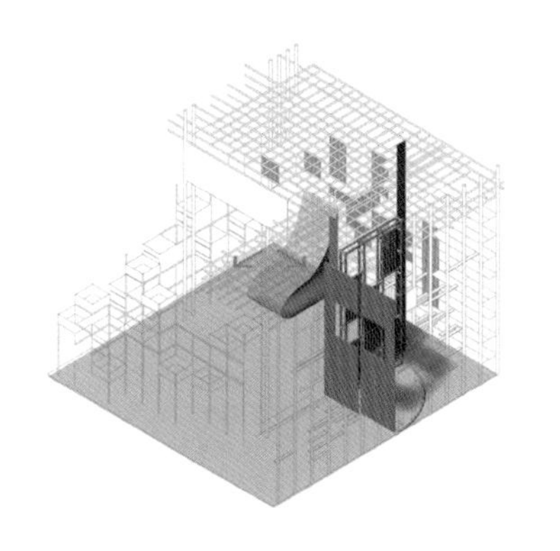

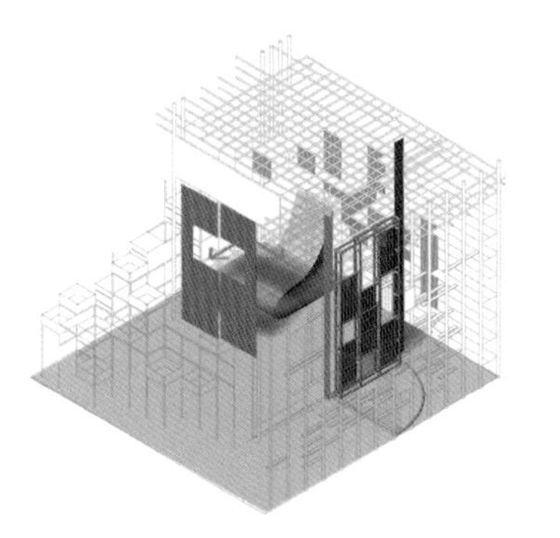

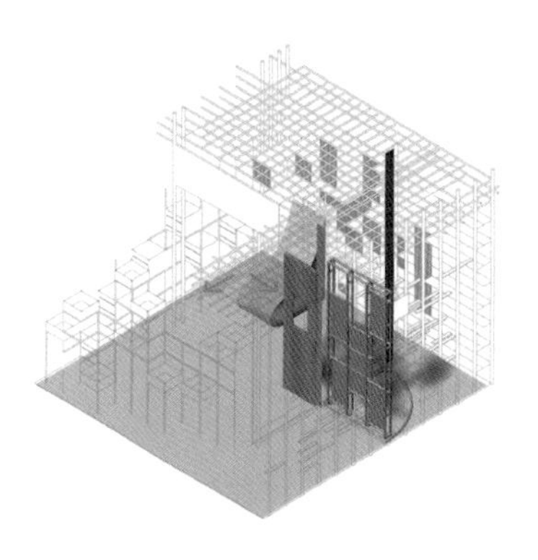

等雨到
WAITING FOR THE RAIN

“泉壑带茅茨，云霞生薜帷。 竹怜新雨后，山爱夕阳时”。这是设计概念形成时脑海中出现的场景。作品探讨叙事型空间营造，将雨作为一种不期而遇的自然要素，融合在设计中，弥补了场地无水景的缺憾，同时强调了场地的情景体验。依据参与者距构筑物距离的由远及近，形成初见—徘徊—彳亍—等待—邂逅五个场景。

构筑物通过四个方向韵律一致的直立竹竿形成信息亭的围合空间，模糊了内外空间的界限，形成了独特的空间体验。结合结构知识形成极简却巧妙的竹排列形态，并形成了丰富的光影变化。

"When the rain falls, the mountain loves sunset." This was the scene in the mind when the design concept was formed. The work explores the narrative space creation, and the rain is a natural element that meets unexpectedly. It is integrated into the design to make up for the shortcomings of the waterless scene of the site, and at the same time emphasizes the experience of the venue. Five scenes are constructed according to the distance between the visitor and the site, which is called first meeting, wandering, wading through, waiting and encountering.

The design constructs the enclosed space through four vertical bamboo poles with consistent prosody, which blurs the boundaries between internal and external space and forms a unique spatial experience. By combining structural knowledge, the design purposes a minimalist but subtle bamboo arrangement with a rich variety of light and shadow.

参赛者学校：北京林业大学
指导老师：王沛永、郑曦
参赛者：高珊、陈希希、楼前、孙瑾玉、田笑常、徐菱励、王诗濛、贾一非
School: Beijing Forestry University
Advisers: Wang Peiyong, Zheng Xi
Participants: Gao Shan, Chen Xixi, Lou Qian, Sun Jinyu, Tian Xiaochang, Xu Lingli, Wang Shiying, Jia Yifei

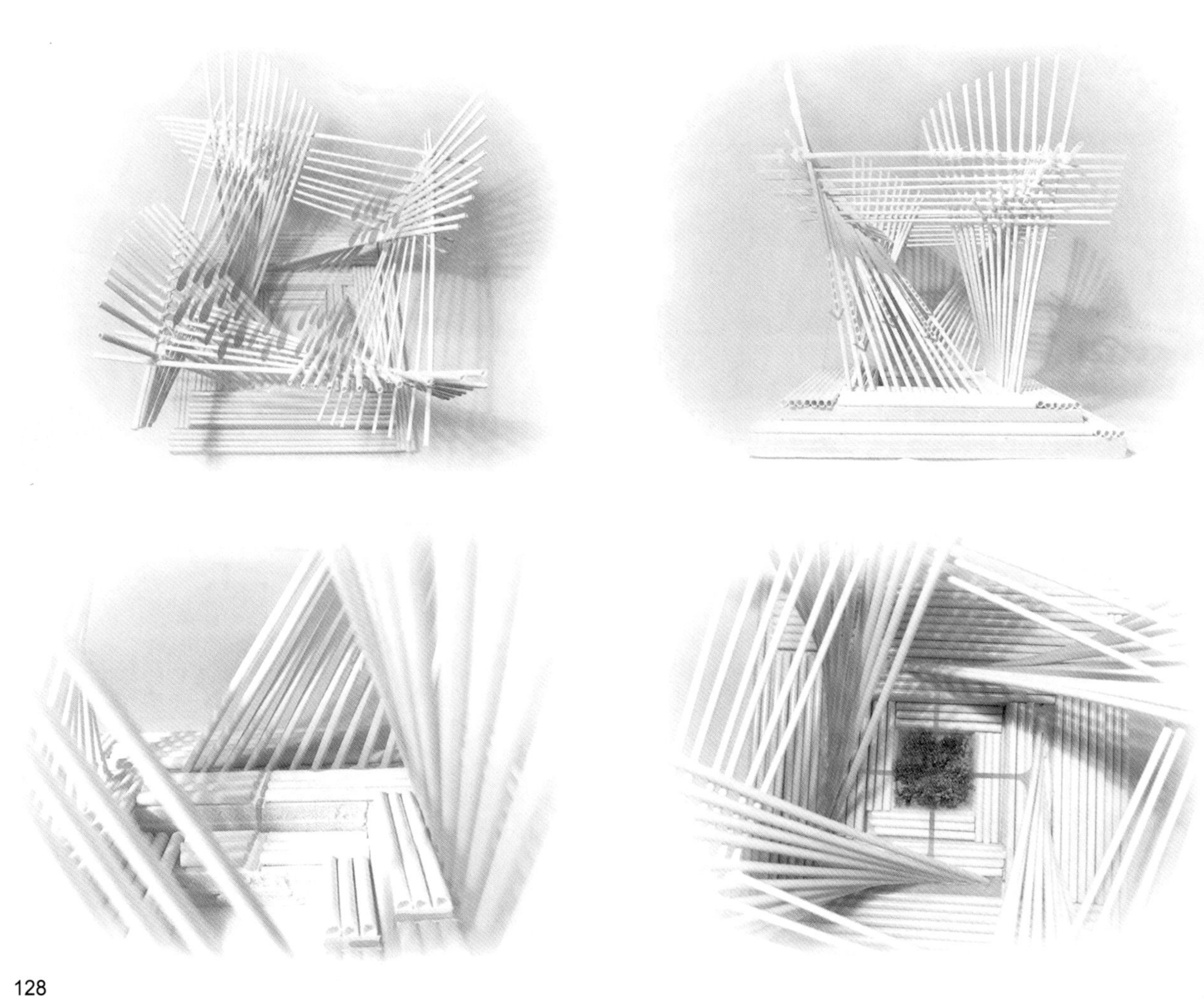

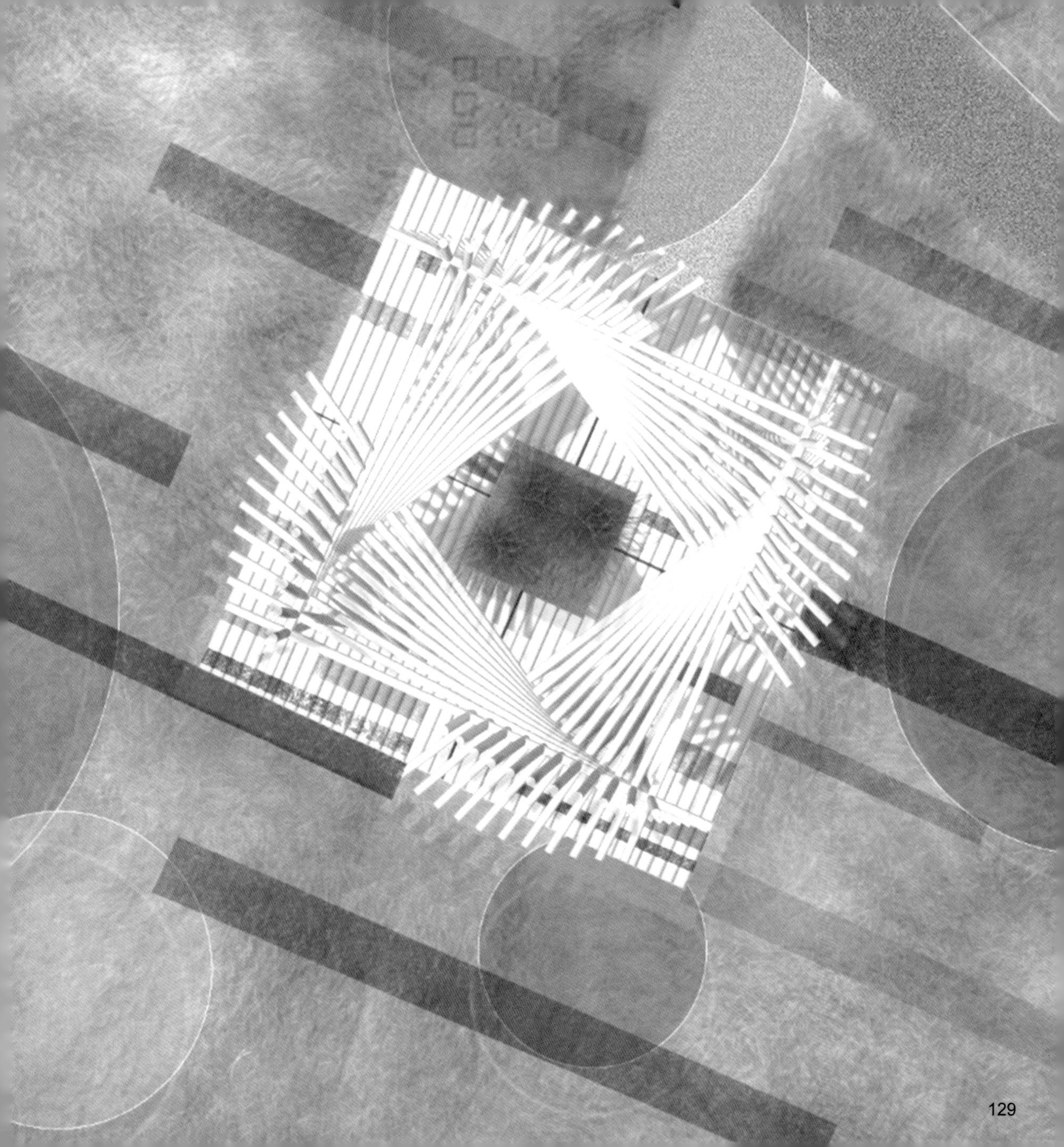

双笙·并蒂
COMMENSALISM & BETTY

竹修长纤细，有“花中君子”之称；花灼灼其华，绽放生的魅力。竹和花同生存、共呼吸，展现着生的两种姿态。竹一曰卓尔，二曰善群。以竹喻花，构筑为亭；以花衬竹，相映成景，风韵多姿，花香清远。

竹亭主要选取了韧性强、可塑性高的黄竹和红竹，并通过捆绑、穿孔等方式进行搭建。

Bamboo is known as the gentleman of the flower and has a slender, long character. the flowers burn, blooming the charm of life. The bamboo and the flowers breathe together with the living, showing the two gestures of life. With bamboo metaphor flowers, architecture for the pavilion, the design is equipped with charm and great fragrance.

The bamboo pavilion mainly chooses yellow bamboo and red bamboo with strong toughness and high plasticity, and is built by binding and perforating.

参赛者学校：北京林业大学
指导老师：公伟
参赛者：向琦、刘思琴、王秋霞、王雅欣
School: Beijing Forestry University
Advisers: Gong Wei
Participants: Xiang Qi, Liu Siqin, Wang Qiuxia, Wang Yaxin

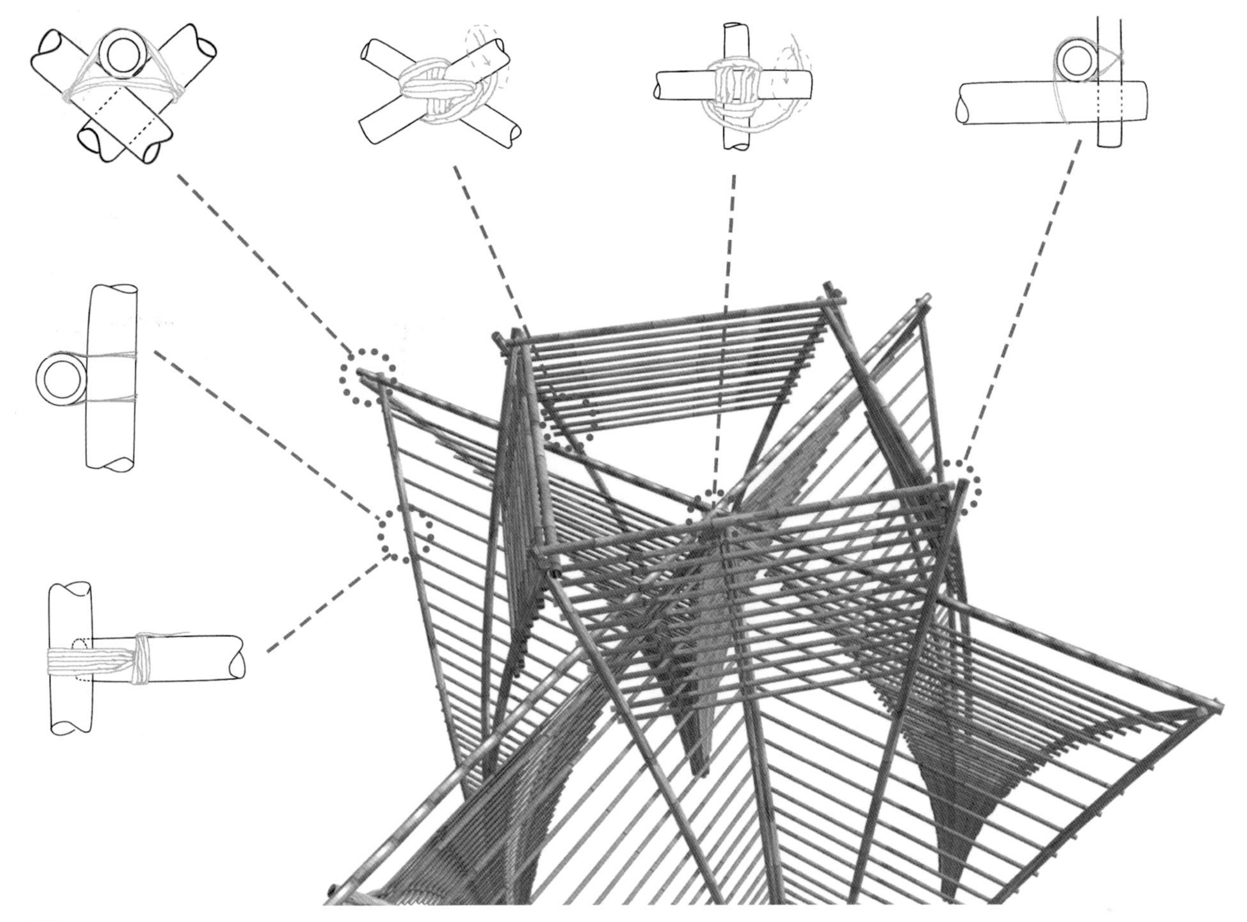

N

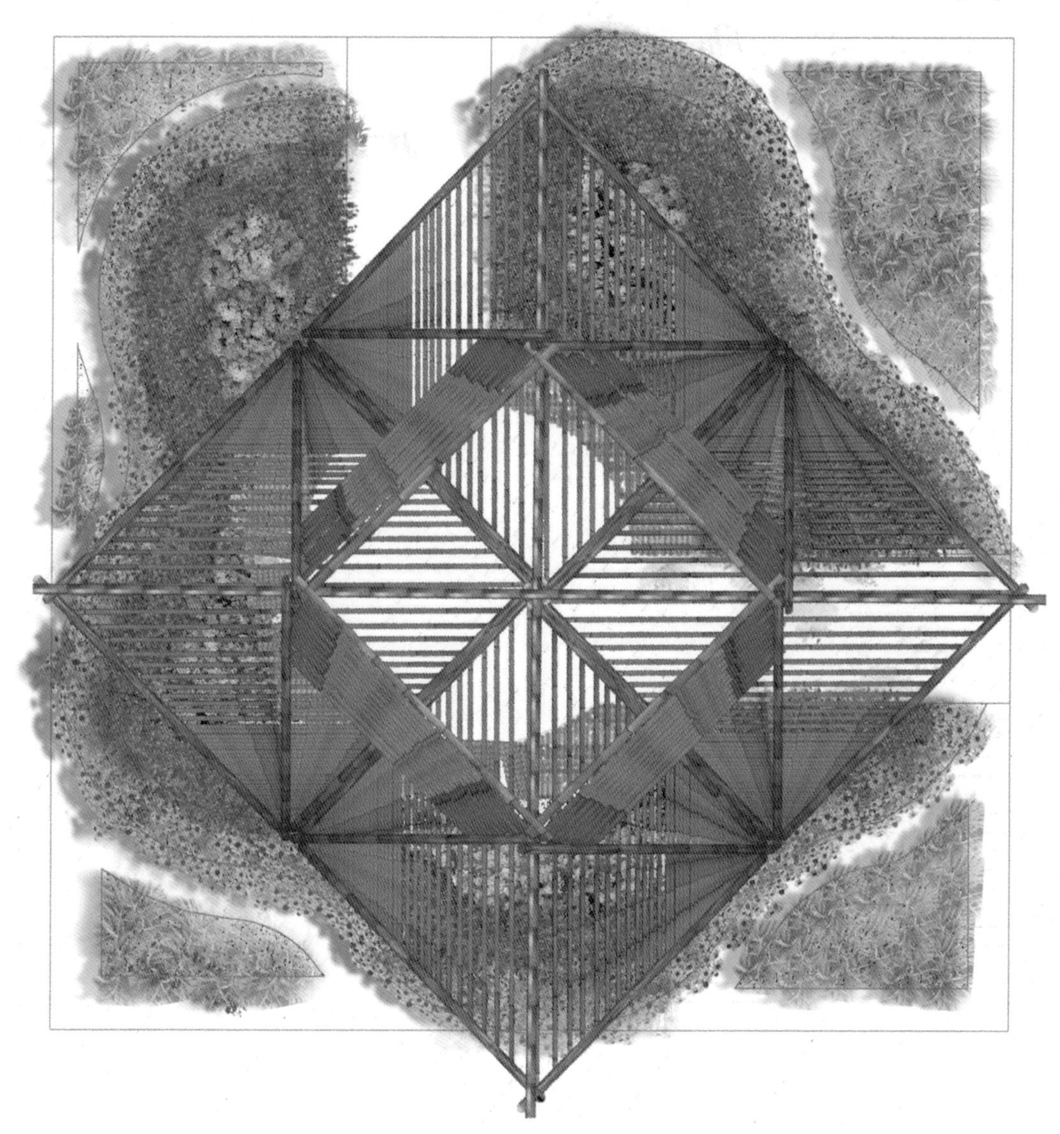

第二届（2019）
北林国际花园建造节

The 2nd (2019) BFU International Garden-Making Festival

主题：
花园的诗意

方案征集时间：
2019 年 3-10 月
实地建造时间：
2019 年 10 月 10-13 日
开放活动时间：
2019 年 10 月 13-20 日

报名选手：
365 组设计团队，来自 103 所高校，共 2004 人
实地建造团队：
入围团队 8 个、受邀参加团队 7 个

主办单位：
北京林业大学、中国风景园林学会教育工作委员会、国际竹藤组织
承办单位：
北京林业大学园林学院、浙江竹境文化旅游发展股份有限公司、《风景园林》杂志
支持单位：
中国建筑学会园林景观分会、北京市花木有限公司

Theme:
Poetic Garden

Call for Entry:
March–October, 2019
On-site Construction:
October 10–13, 2019
Opening:
October 13–20, 2019

Registered Teams:
365 teams, from 103 universities, a total of 2,004 students
Construction Teams:
8 shortlisted teams and 7 invited teams

Organizers:
Beijing Forestry University, Chinese Society of Landscape Architecture Education Committee, International Bamboo and Rattan Organization
Executive Organizers:
School of Landscape Architecture of Beijing Forestry University, Zhejiang Zhujing Cultural Tourism Development Co., Ltd., Landscape Architecture Journal
Support Organizers:
Institute of Landscape Architecture of the Architectural Society of China, Beijing Flowerscape Co., Ltd.

樊笼

FUN CAGE

心为形役，尘世马牛，身被名牵，樊笼鸡鹜。生活犹如围城，里面的人想出来，外面的人也想进去。

树叶梳篦的光斑把竹藤的影子一寸一寸彼此递交，给这座林下的樊笼增添了古朴野趣，令人心生向往。

这座樊笼亦如围城，向外拥抱自然，消化于广袤和汪洋；向内窥得自然的方寸之美，落入零星的萤火。
笼里笼外，哪边才是真正的自然?
正如陶渊明诗中所道“久在樊笼里，复得返自然”。
只有穿梭于樊笼内外，体验过视角的转移、心境的不同，才能悟出何谓樊笼以及自然本意。

请君入樊笼，体验在个中。

If the mind is dominated by fame and fortune, we will live in the world like cattle and horses. If the body is bound by fame, we will lose freedom like a caged chicken. Life is like a besieged city. Those inside want to get out, meanwhile, those outside want to get in.

The light of the leaves and the shadow of the bamboo cane inch by inch intertwined. The desirable scene added to the primitive and wild interest of the enclosure under the forest.

Like a walled city, this fun cage embraces nature outwards and digests itself in the vastness as well as peeps inwards into the small beauty of nature and the scattered fireflies.
So inside and outside the cage, which side is the real nature?
As Tao Yuanming said in his poem, "Long in the cage, return to nature."
It is only when we travel inside and outside the cage, experience the shift of perspective and the change of state of mind that we can understand the meaning of the cage and nature.

Please enter the cage and experience in it.

参赛者学校：北京林业大学
指导老师：郑小东、李慧
参赛者：童佳玉、余泳欣、何黛娜、张娜、蔡奇峰、梁金子、刘汇子、朱梓宁
School: Beijing Forestry University
Advisers: Zheng Xiaodong, Li Hui
Participants: Tong Jiayu, Yu Yongxin, He Daina, Zhang Na, Cai Qifeng, Liang Jinzi, Liu Huizi, Zhu Zining

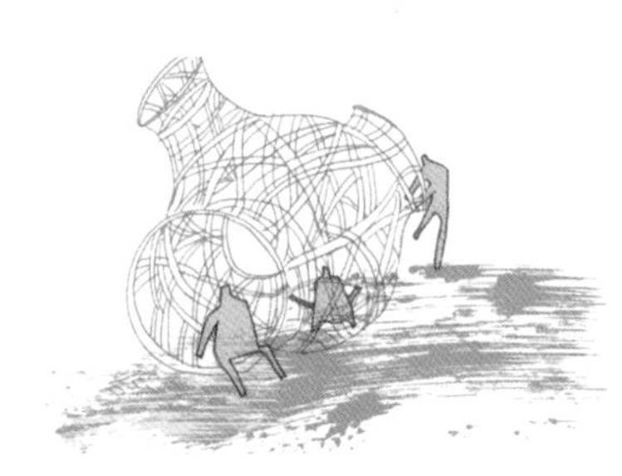

人群活动 /People activities

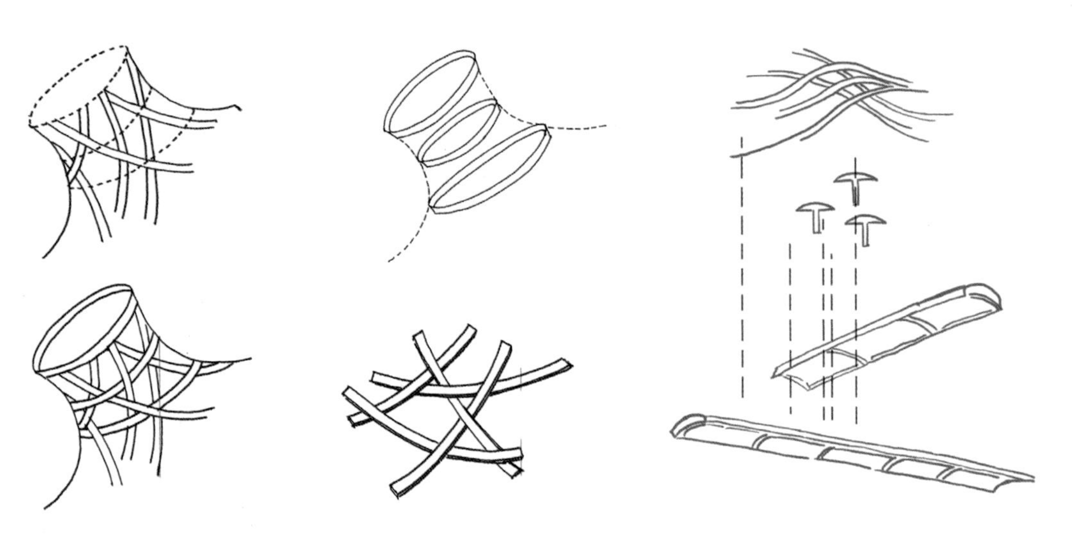

结构 /Structure

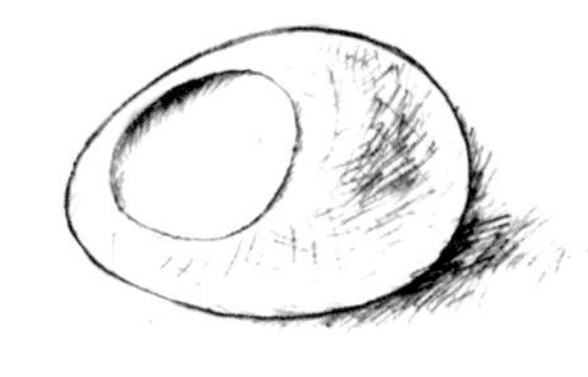

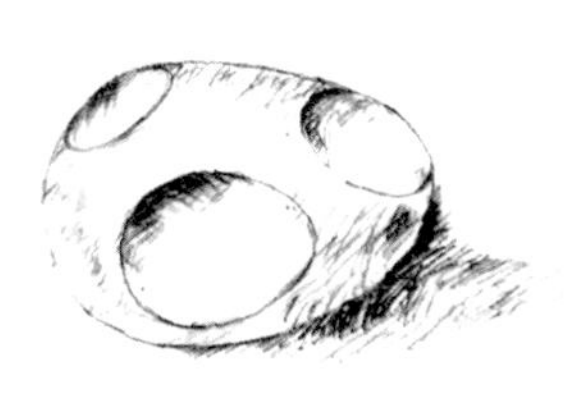

概念 /Concept

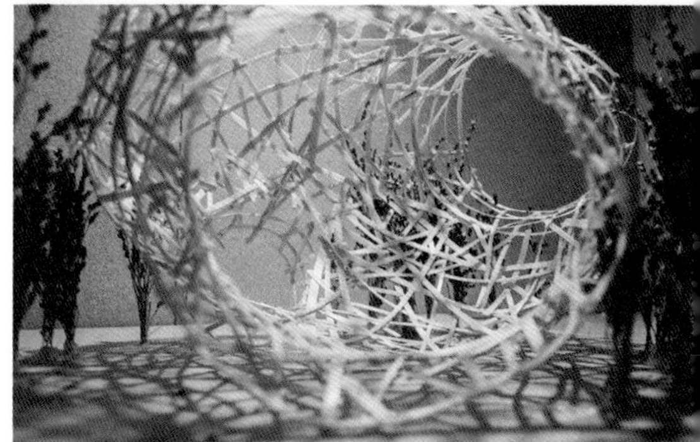

无序之序
BETWEEN THE ORDERS AND CHAOS

“混沌之先，太无空焉。混沌之始，太和寄焉……洎乎元气蒙鸿，萌芽兹始，遂分天地，肇立乾坤，启阴感阳，分布元气，乃孕中和……成天立地，化造万物。”

世界万物由无序中诞生，又偏爱追寻有序，最终消亡，回归无序。有序与无序、天与地、阴与阳总是在动态中取得平衡。本作品以竹节为地，象征有序；以竹片为天，象征无序。竹片与竹节，本是一体，暗合阴阳造化轮回不息之理。

"At the beginning of chaos, there was nothing in the world. Slowly peace appeared, but it was very not obvious. Then a hazy, endless mass of air was in constant motion, from which everything sprouted. Then the heaven and earth began to separate. The positive materiality becomes *qi*, rising and then becoming the so-called blue sky. The negative material continuously condensed and becoming the vast earth. The *qi* of *yin* and *yang* were constantly in motion, canceling each other out and becoming the heaven and earth, giving birth to all things."

Everything in the world was born out of chaos, but it always pursued the order. Eventually, everything will die and return to chaos. Chaos and disorder, heaven and earth, *yin* and *yang* always balance in dynamics. This work takes bamboo tube as the ground which symbolizes order. And it uses the bamboo strip as the sky that expresses disorder. The whole work conceals the *yin* and *yang* of the reincarnation of the cycle.

参赛者学校：日本国立千叶大学
指导老师：三谷徹、霜田亮祐
特别指导：章俊华
参赛者：山﨑祥平、水谷苍、山下雅弘、刘书昊、鲤川哲平、佐佐木圭、谷本实有、严妮
School: Chiba University, Japan
Advisers: Mitani Toru, Shimoda Ryosuke
Special Advisor: Zhang Junhua
Participants: Sakai Takahiro, Mizutani So, Sasaki Kei, Yamashita Masahiro, LiuShuhao, Koikawa Teppei, Tanimoto Miyu, Yan Ni

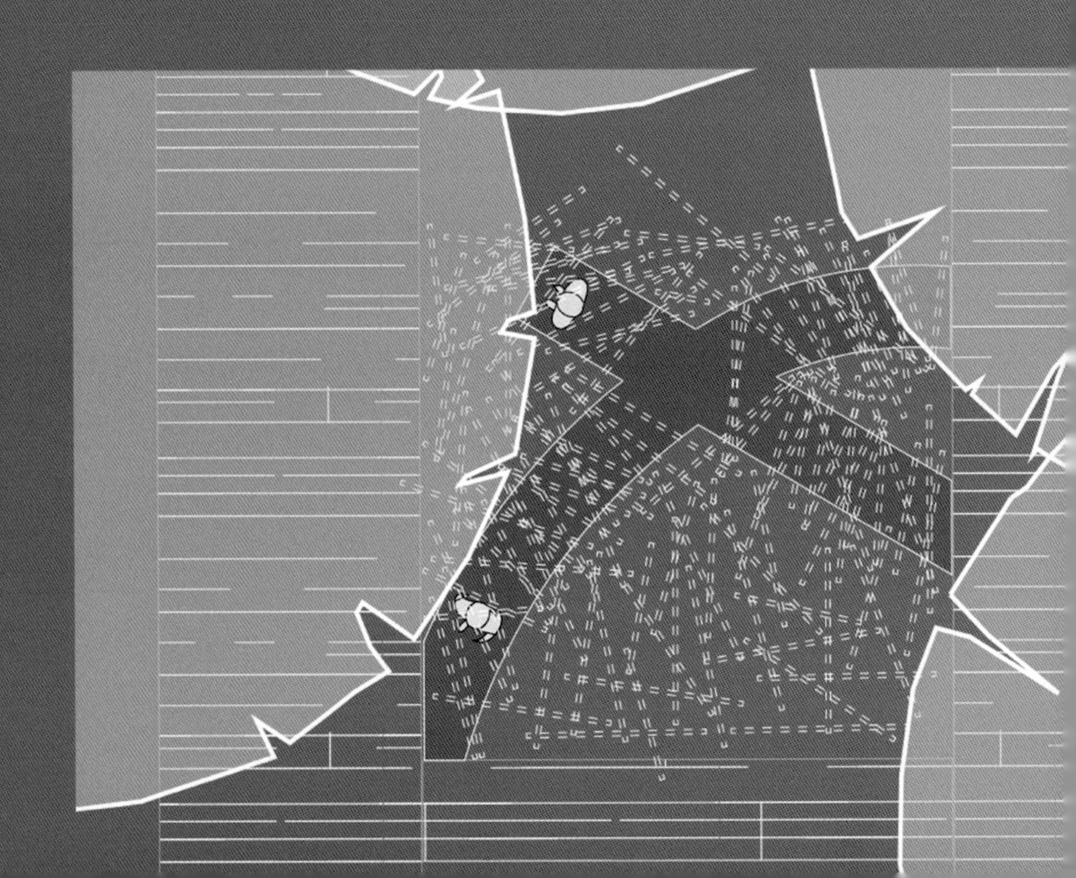

弹弹弹
CRUNCH!

生活使人小心翼翼，
我们想轻快地逃离，
轻盈地走在云朵上，
温柔地俯视这世界。

生活充满着紧张与忙碌，充斥着微妙的控制与失控。花园设计灵感来源于竹子富有弹性的特征，希望以“弹”为感受的核心，在人们活动的过程中产生一连串不断变化的视觉与空间体验。

是压迫感、小心翼翼、紧张与不安？
还是轻盈的起伏，身侧光影错落，思绪随之浮动？
当我们重新回到地面，花园再次出现在眼前，但轻微的晃动感依旧存在，周围的世界，还是原先那一个吗？

Life makes one cautious. We want to take a light escape from life, walk lightly on the clouds and look down gently on the world.

Life is full of tension and busyness, subtle control and loss of control. Inspired by the elastic characteristics of bamboo, the design of garden aims to create a series of changing visual and spatial experiences with "bounce" as the core of feeling.

Will these experiences be stressful, cautious, tense and unsettling, or undulation with thoughts floating accordingly, accompanying with light and shadow strewn at random?
When we return to the surface, the garden reappears. But is the world around the same as before with the slight shaking still there?

参赛者学校：华南理工大学
指导老师：林广思、熊璐
参赛者：王曲荷、刘瑶、钟和丽、李悦、茌文秀、沈攀、陈梦芸、周兆森
School: South China University of Technology
Advisers: Lin Guangsi, Xiong lu
Participants: Wang Quhe, Liu Yao, Zhong Heli, Li Yue, Chi Wenxiu, Shen Pan, Chen Mengyun, Zhou Zhaosen

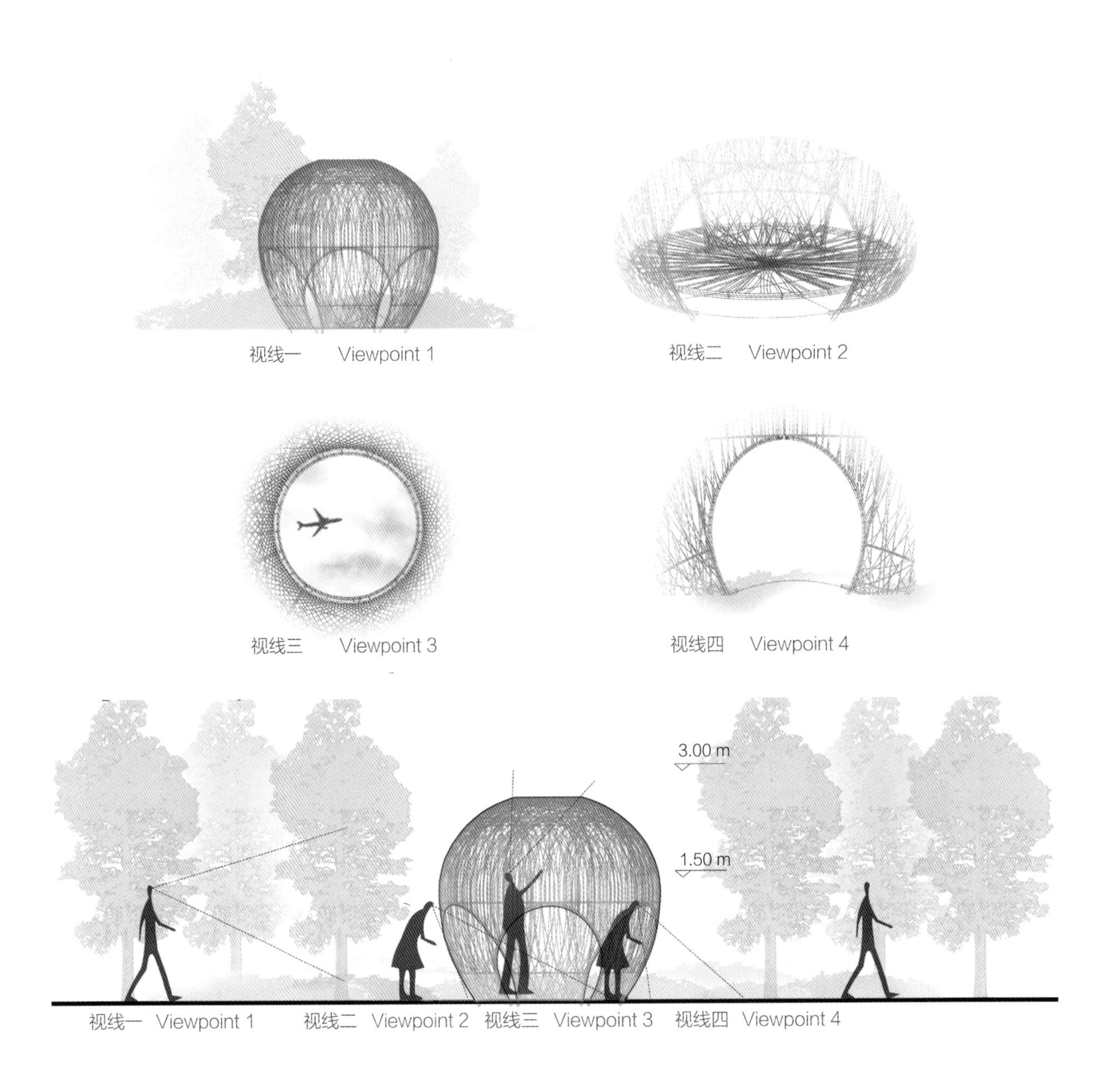
视线一 Viewpoint 1
视线二 Viewpoint 2
视线三 Viewpoint 3
视线四 Viewpoint 4
3.00 m
1.50 m
视线一 Viewpoint 1
视线二 Viewpoint 2
视线三 Viewpoint 3
视线四 Viewpoint 4

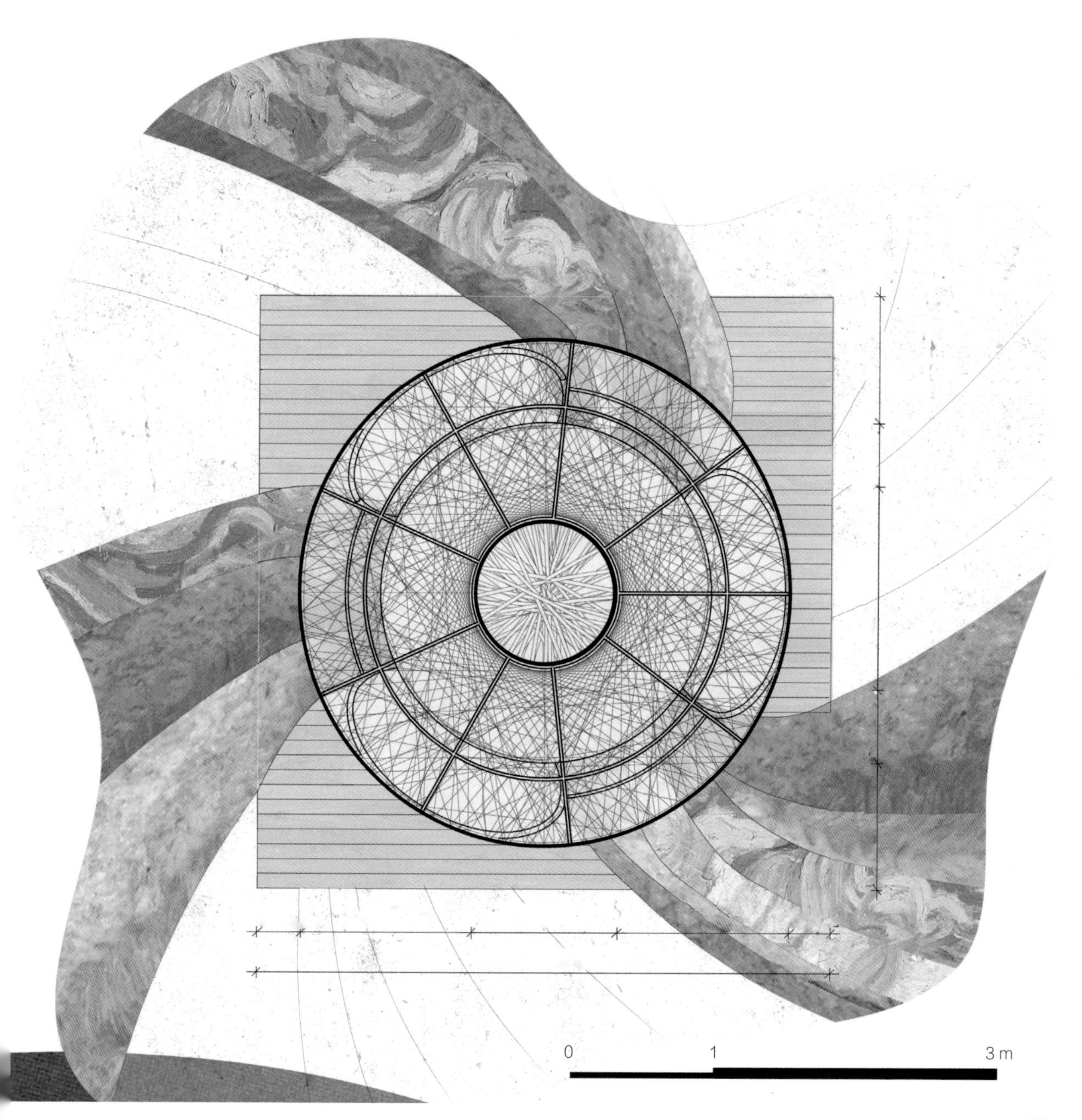
0
1
3 m

那伽：启蒙和觉悟的重生

NAGA : THE REBIRTH TO ENLIGHTENMENT

这个展览馆代表着那伽的居所。那伽是一种可以变成人形的巨蛇，所以常被误以为是佛教小僧侣。佛陀召唤了那伽并告诉它不要再变成僧侣的模样，这让那伽万般沮丧然后开始哭泣。这巨蛇被赋予了五条戒律作为它来生转世为人的条件，然后它便能选择是否成为一名真正的僧侣。如果破了这些戒律，则必须退出。

园林景观的设计概念就像达到启蒙和觉悟的方式，佛陀将人类比喻成四种莲花的象征：

一是泥下的莲花，它不会理解任何事物；

二是水下的莲花，它必须花时间去理解某些事物；

三是灿烂的莲花，它能够快速地理解任何事物，但它们需要来自各处的指导和知识；

四是真正明亮和绽放的莲花；它可以瞬间并正确地理解一切。

其实佛教的启蒙与觉悟是需要结合正念和了解并停止欲望的生源。当人们平静或 dhyāna（梵语）或 jhāna（巴利语）时，他们会领悟一切。

This pavilion represents the Naga's hut. Naga is a giant snake that can take the form of a human, so it is often mistaken for a young Buddhist monk. The Buddha summoned Naga and told him not to turn into a monk again, which made Naga depressed and began to cry. The serpent was given five commandments as a condition for its reincarnation as a human in its next life. If he fulfills these precepts, he can choose whether to become a true monk or not. If these commandments are broken, he must withdraw.

The concept of landscape design is like a way of enlightenment. The Buddha likened mankind to four lotus symbols:

The first is the lotus under the mud. It does not understand anything.

The second is the underwater lotus which has to take time to understand something.

The third is the brilliant lotus which can understand anything quickly but needs guidance and knowledge from everywhere.

The fourth, a truly bright and blooming lotus flower. It can understand everything instantaneously and correctly.

In fact, enlightenment in Buddhism is the source of a combination of mindfulness, understanding and cessation of desire. When people are calm, they understand everything.

参赛者学校：泰国朱拉隆功大学

指导老师：Terdsak Tachakitkachom、Ariya Aruninta

参赛者：Svit Piriyasurawong、Theappitaknillawan、Sirirak Wongsim、Komkanokrit Chauprampare、Natdanai Tumpanuwat、Sviadol Footrakui

School: Chulalongkorn University, Thailand

Advisers: Terdsak Tachakitkachom, Ariya Aruninta

Participants: Svit Piriyasurawong, Theappitaknillawan, Sirirak Wongsim, Komkanokrit Chauprampare, Natdanai Tumpanuwat, Sviadol Footrakui

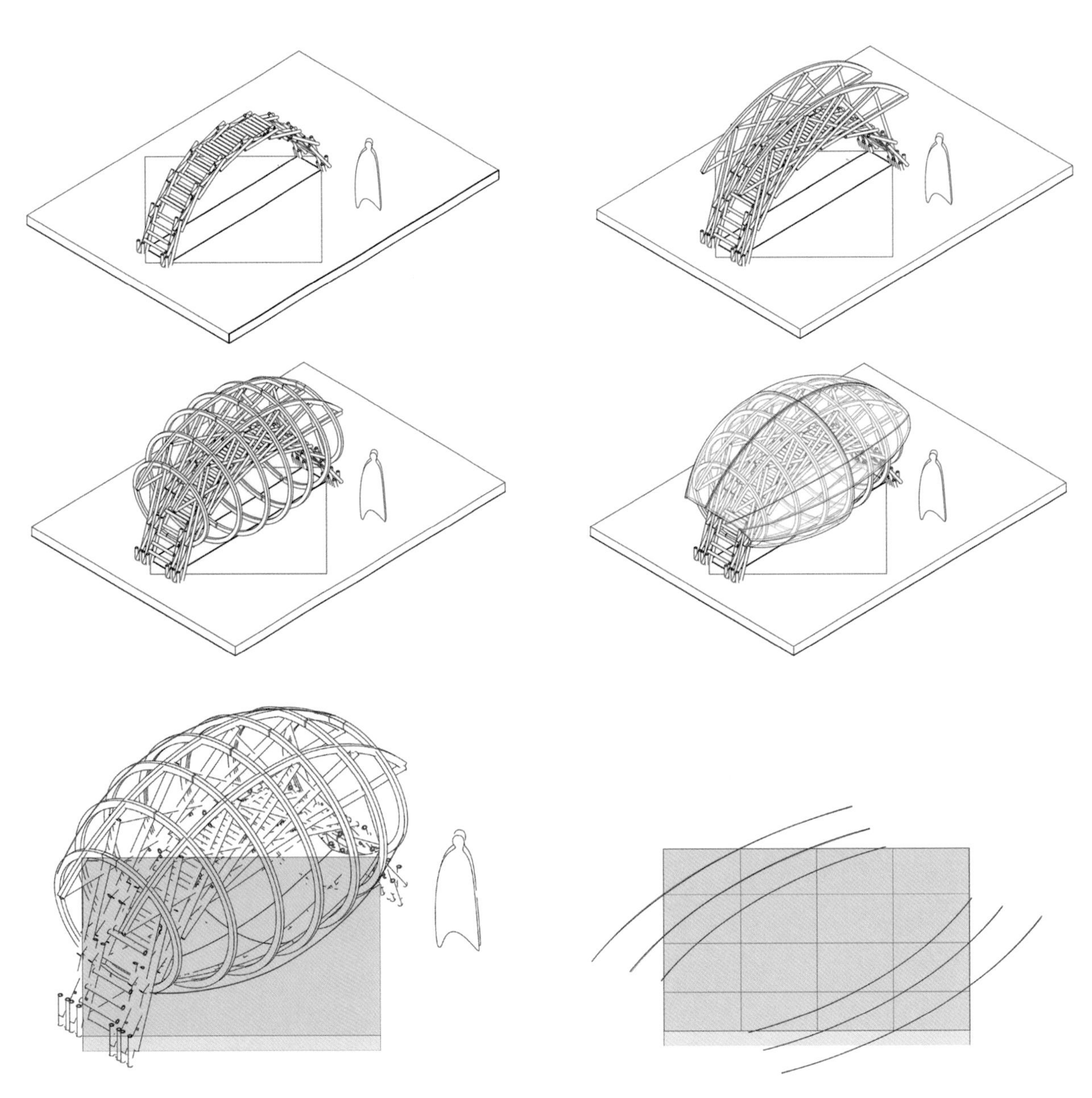

设计基础 /Base of design

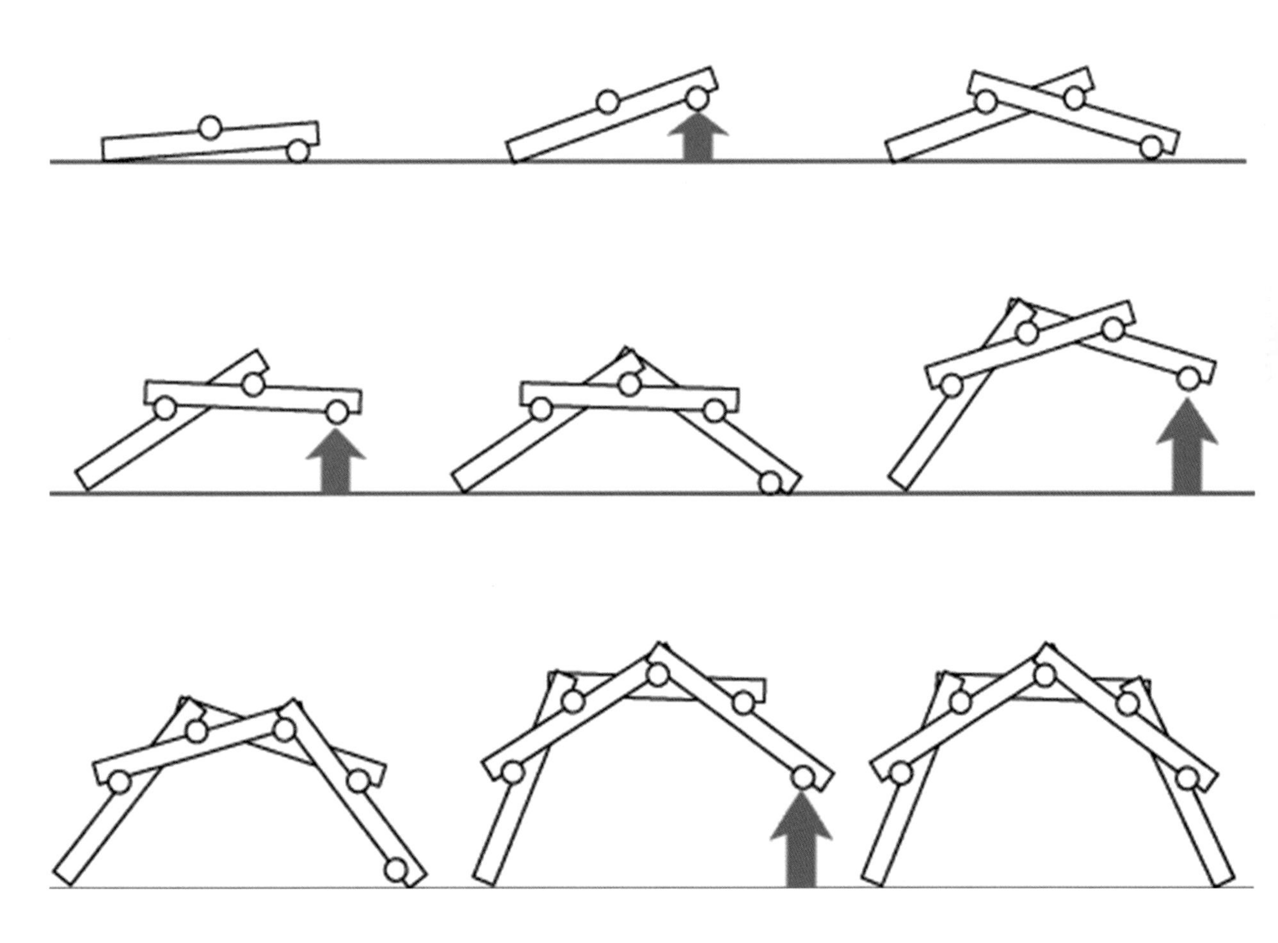

筠林·见园
GARDEN APPEARED IN BAMBOO FOREST

方案“筠林·见园”，用隐喻、象征、正负形的转换，塑造隐藏在竹林中的六角小亭（负形），亭周围假山错落，竹树浓密，宛如优美的田园诗。亭是中国古典园林的重要构成元素，体现了天人合一、人与自然和谐的理念。

设计强调朦胧美和意境美。在 4m×4m×4m 的空间内，用亭的实形对密布的细竹竿进行切割，得到了亭子的负形。从外面看，亭的形状若隐若现；进入亭内，又仿佛置身茂密的竹林之中。

设计难点在于如何切割形体，让负形准确，又不生硬。因此选用了六角亭，轮廓简化为三角和六边形底座，沿斜角切割，明确了入口的同时，形态又不显得生硬。构件经过多次简化，尽量保证施工的模块化和简便化。同时，构件也考虑了一定的伸缩性，适应竹子直径间有差别的特点。

The project "Garden Appeared in Bamboo Forest" uses metaphors, symbols and the transformation of positive and negative forms to create hexagonal pavilions hidden in the bamboo forest. Pavilion is around the scattered rockery and thick bamboo trees, which likes a beautiful idyllic. Pavilion is an important component of Chinese classical garden, which embodies the idea of harmony between man and nature.

The design emphasizes hazy beauty and artistic conception beauty. In the space whose size is 4 m × 4 m × 4 m, the dense thin bamboo rods were cut with the shape of the pavilion and the negative shape of the pavilion was obtained. From the outside, the shape of the pavilion is looming. When one enters the pavilion, it seems to be in the thick bamboo forest.

The difficulty of design is how to cut the shape, so that the negative shape is accurate instead of stiff. Therefore, the hexagonal pavilion was chosen. The outline is simplified to triangular and hexagonal bases. At the same time, the contour is cut along the bevel Angle so that the entrance is clear as well as the shape is not stiff. The components have been simplified for many times to ensure the modularization and simplicity of construction. The structure is also considered to be flexible to accommodate the difference between the diameters of bamboo.

参赛者学校：中央美术学院
指导老师：吴祥艳、张茜
参赛者：胡宇晨、谢秀竹、孙昆仑、赵文健、王薪宇、陈宁、吴霁、廉景森
School: Central Academy of Fine Arts
Adviser: Wu Xiangyan, Zhang Xi
Participants: Hu Yuchen, Xie Xiuzhu, Sun Kunlun, Zhao Wenjian, Wang Xinyu, Chen Ning, Wu Ji, Lian Jingsen

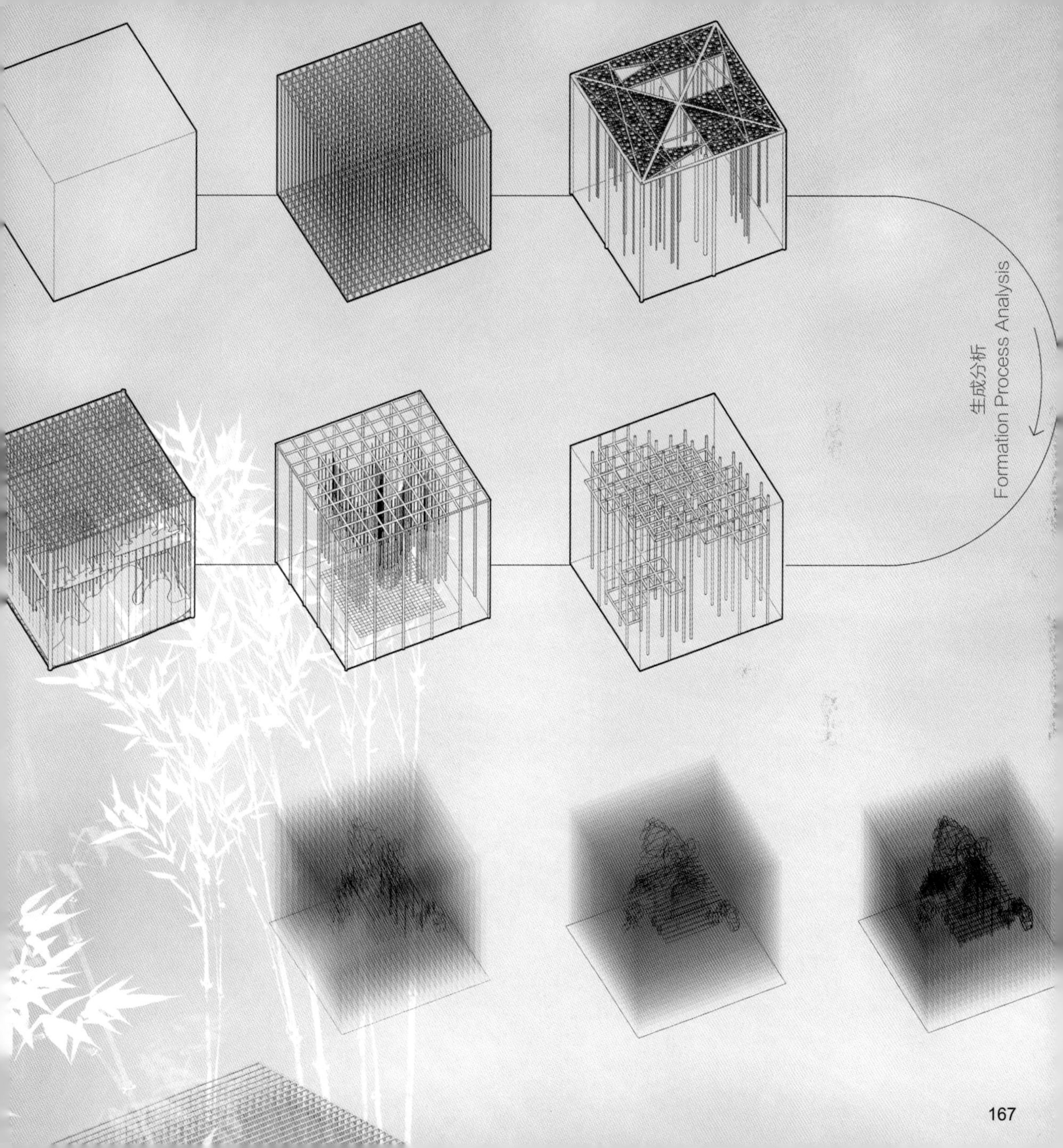
生成分析
Formation Process Analysis

黑洞
BLACK HOLE

宇宙是一座相对真实与绝对虚幻错乱交织的花园，而人们对宇宙的探索永无止境，探索得越多，越接近真相，越接近它唯一的尽头。

方案受首次可见的“黑洞”所启发，由三个各自倾斜的椭圆平面相互交错形成横向骨架，纵向为交叉编织的弧形竹篾。顶部的“洞”象征着无限延伸的宇宙，但“宇宙”并非遥不可及。中央倒挂的竹帘仿佛“黑洞”一般有着超强的磁力，将人们吸入其中，为沉思提供了更多的空间。被“拨开”的一角成为“黑洞”的入口，白色的碎石铺地营造出静谧的氛围，四周高大密植的狼尾草围合成隐秘而烂漫的沉思空间。

观者席地而坐，在烂漫的花园中，抬头纵观广阔的天空，已然坠入宇宙与自我的遐想。

The universe is a garden that is intertwined with real and absolute illusions. The romantic exploration of humans in universe is endless. The more we explore, the closer we get to the truth which near the unique end.

Inspired by the black hole, the scheme is formed by three respectively tilted elliptical planes to form a horizontal skeleton, curved bamboo cross-weaving in vertically. The "hole" at the top symbolizes an infinitely extended universe, but the "universe" is not out of reach. The bamboo curtains hanging in the center is just like the "black hole" with a super magnetic force, attracting people into for wild imagination. The "pick-up" corner becomes the entrance to the "black hole". The white gravel pavement creates quiet atmosphere and the tall, dense pennisetum surrounded creating a secret, romantic space for thinking.

The viewer sits on the ground, looking up at the vast sky and then gradually fall into the wild imagination about the universe and ego.

参赛者学校：北京林业大学
指导老师：冯潇、段威
参赛者：钱小琴、卢靖、谢家琪、李婷、廖菁菁、聂蕾、梁彤、杨瑞莹
School: Beijing Forestry University
Advisers: Feng Xiao, Duan Wei
Participants: Qian Xiaoqin, Lu Jing, Xie Jiaqi, Li Ting, Liao Jingjing, Nie Lei, Liang Tong, Yang Ruiying

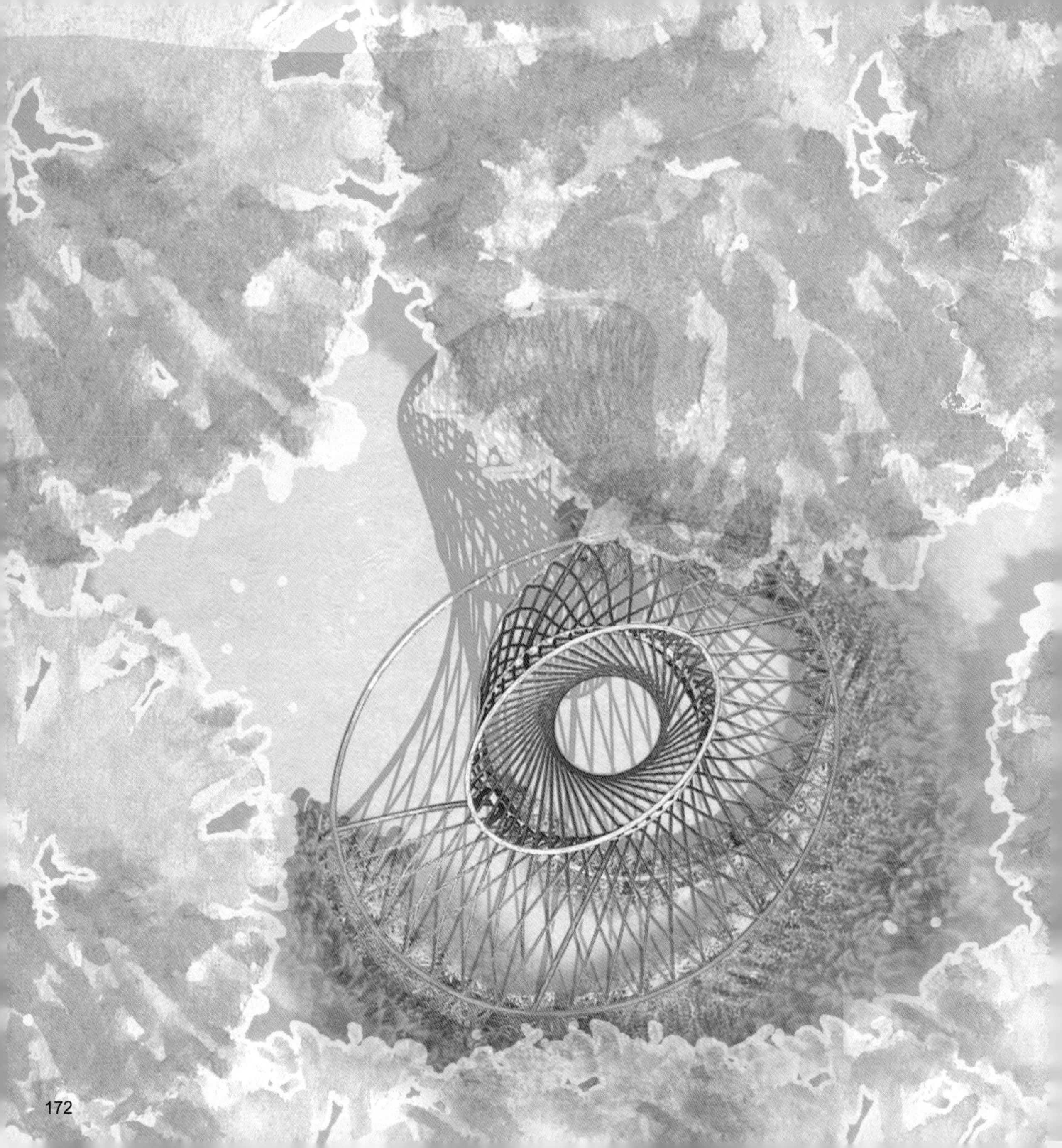

起风了
THE WIND RISES

风，能使晚秋的树叶脱落，能催开早春二月的鲜花，刮进竹林时可把万棵翠竹吹得歪歪斜斜。风轻轻起，竹建筑宛如花园里少女的裙摆，又如随风飞舞的落叶，抑或是翩翩起舞的蝴蝶。阳光透过竹篾的缝隙，洒满全身，本方案希望营造一个可以供人休憩、娱乐、冥想等的交流空间，在校园中提供一处融入大自然的驻足点。

The wind can make the leaves of the late autumn fall off and promote the flowers of the early spring. When the wind scrapes into the bamboo forest, it blows the bamboos swing. The wind rises gently, making the bamboo architecture like a girl's skirt in the garden, a leaf flying with the wind or a dancing butterfly. The sun shines through the gaps of bamboo rafts. This scheme wants to create a communication space for people to rest, entertain, meditate and more as well as provide a stopover for nature in the campus.

参赛者学校：哈尔滨工业大学
指导老师：陆诗亮、余洋
参赛者：周栩至、哈虹竹、于博、蒋雨芊、曾敬、张琦瑀、钱豪杰、罗天一
School: Harbin Institute of Technology
Advisers: Lu Shiliang, Yu Yang
Participants: Zhou Xuzhi, Ha Hongzhu, Yu Bo, Jiang Yuqian, Zeng jing, Zhang Qiyu, Qian Haojie, Luo Tianyi

起风了
The Wind Rises
哈尔滨工业大学

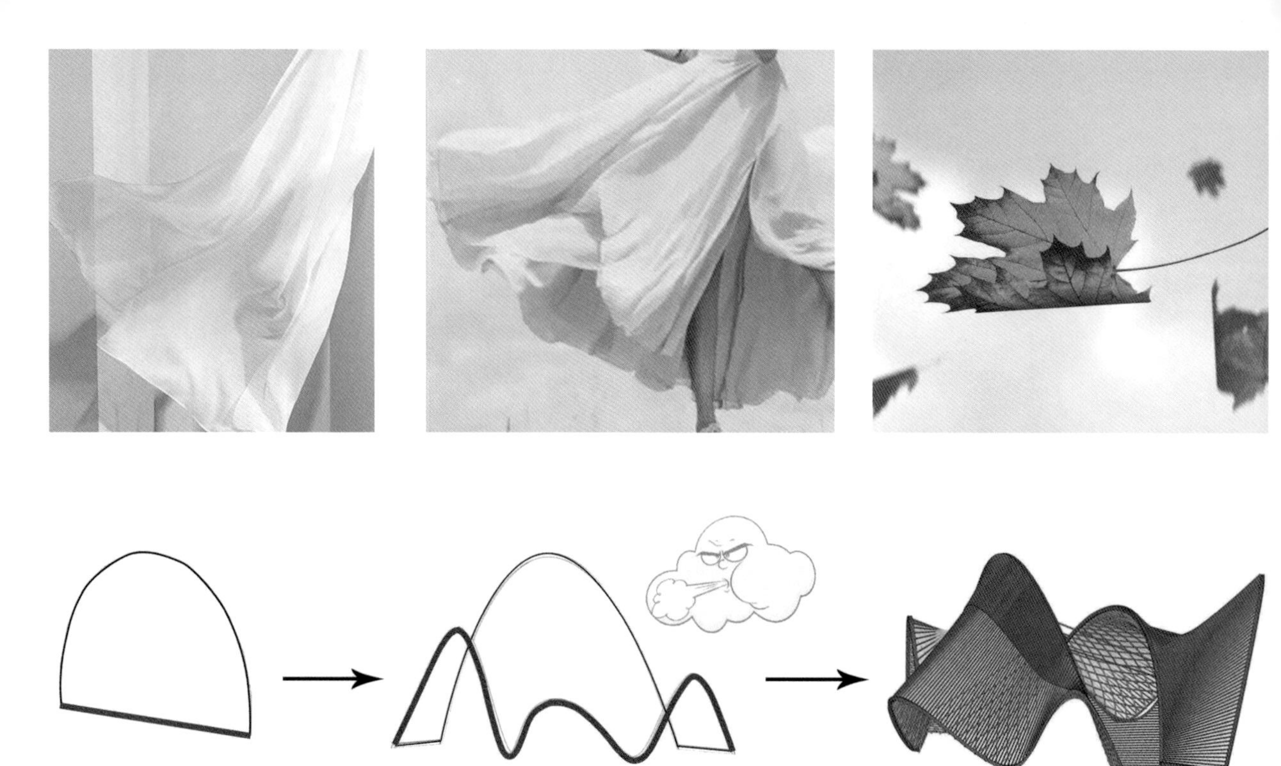

概念生成 /Concept generating

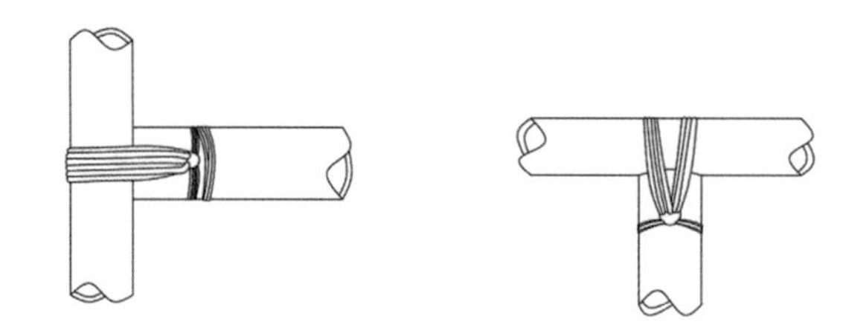

节点构造 /Node construction

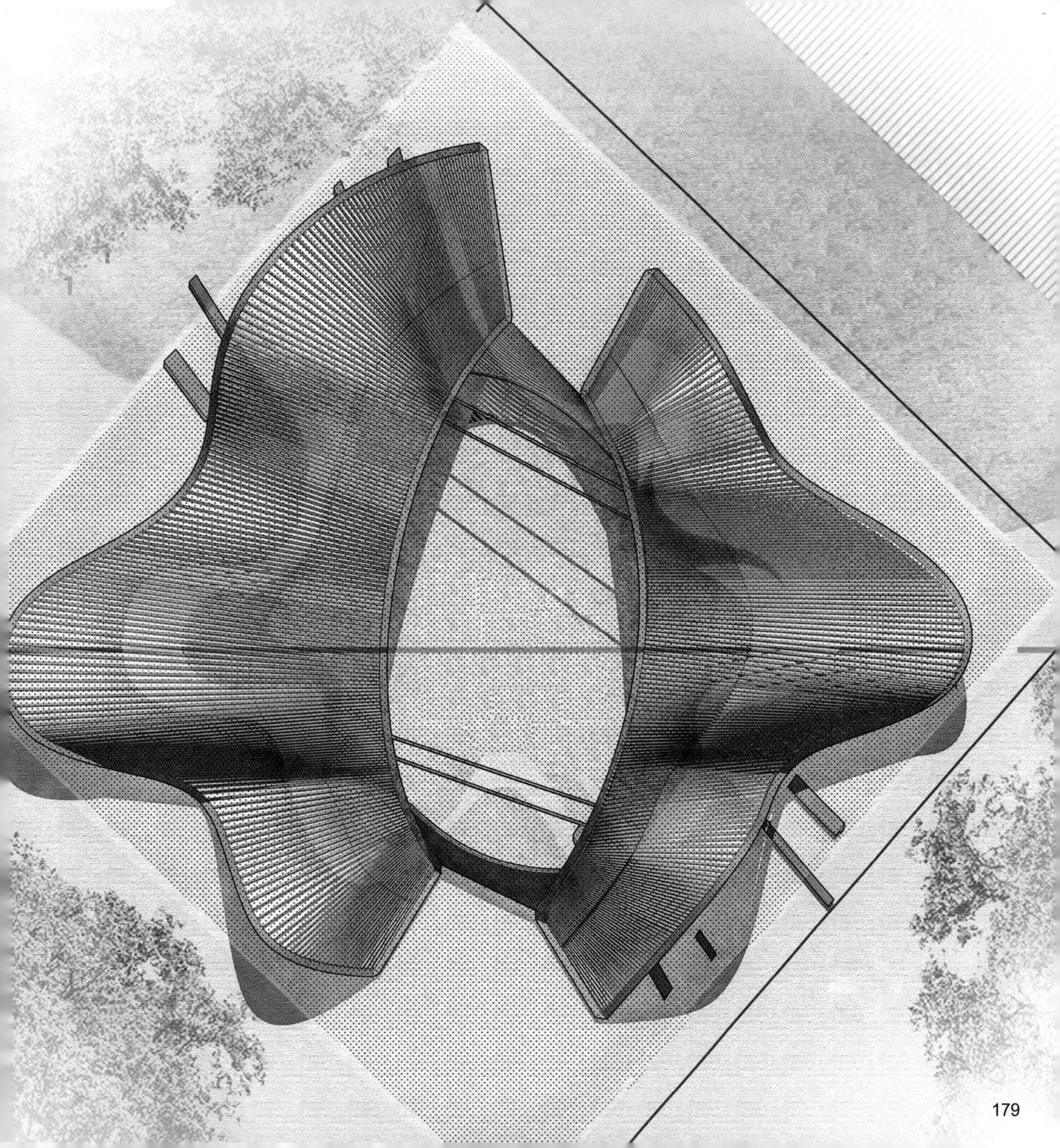

起风了

驰隙

CELEBRATING THE OTHERNESS

城市空隙，又称城市负空间，是广泛存在于街区和建筑物之间的碎片化剩余空间。城市空隙具有灵活性、短暂性和变动性，它不仅是包含替代功能的非程序化空间，也是人们可以自由定义的短暂对象。

为了突出城市空隙的积极表现和生态能力，本方案旨在设计一个小型展馆，通过这个展馆将城市空隙作为一个对象而非空间进行阐释。在这里，竹子和草本植物将根据他们的表现性能被重新定义：作为一个多变的过程，该展馆见证了光影的变化，捕捉植物生长、蔓延、竞争和衰败的全过程。

该展馆将基于主体结构投射形成的花园进行概念表现，运用光影的变化和植物群落逐渐演变下的边缘模糊为主要表现手法。展馆不断变化的每一帧都将体现出城市空隙短暂灵活的概念，让人们重新看到无形的缝隙，并引发参观者对于城市负空间潜在可能性的思考。

Urban void, also called negative space, is the left-over space that existing between gaps of buildings and blocks. Urban void is ephemeral, flexible and transformative. It is not only a non-programmatic space with alternative functions, but also a transient object that people can define freely.

In order to highlight the positive expression and ecological capacity of the urban void, the project aims to design a small pavilion. This pavilion interprets the urban void as an object rather than a space. Here, bamboo and herbs will be redefined according to their performance properties. As a fluid process, the pavilion witnesses the change of light and shade as well as captures the growth, spread, competition and decay of plants.

The pavilion conceptualizes a garden based on the projection of the main structure. The change of light and shadow and the blurring of the edge under the gradual evolution of the plant community are the main techniques of expression. Each changing frame of the pavilion will reflect the concept of the transience and flexibility of the urban gap, allowing people to see the invisible gap again and provoking them to think about the potential of negative space in the city.

参赛者学校：皇家墨尔本理工大学
指导老师：李紫暄、Jock Gilbert
参赛者：诸葛桦莹、徐遥、胡瑾榆、孙铭泽、肖佳丰、许铮
School: Royal Melbourne Institute of Technology, Australia
Advisers: Li Liz, Jock Gilbert
Participants: Zhuge Huaying, Xu Yao, Hu Jinyu, Sun Mingze, Xiao Jiafeng, Xu Zheng

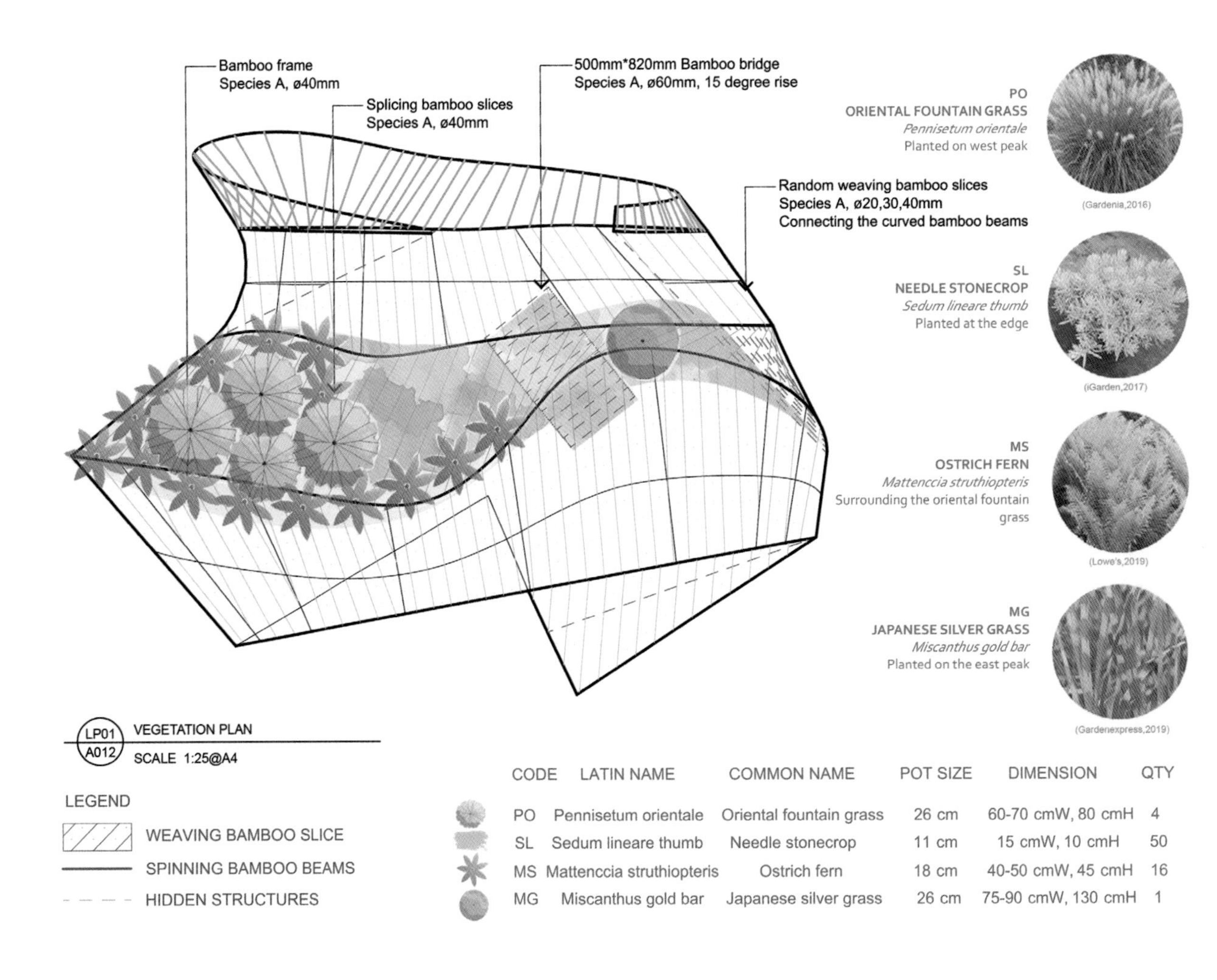

CODE	LATIN NAME	COMMON NAME	POT SIZE	DIMENSION	QTY
PO	Pennisetum orientale	Oriental fountain grass	26 cm	60-70 cmW, 80 cmH	4
SL	Sedum lineare thumb	Needle stonecrop	11 cm	15 cmW, 10 cmH	50
MS	Matterccia struthiopteris	Ostrich fern	18 cm	40-50 cmW, 45 cmH	16
MG	Miscanthus gold bar	Japanese silver grass	26 cm	75-90 cmW, 130 cmH	1

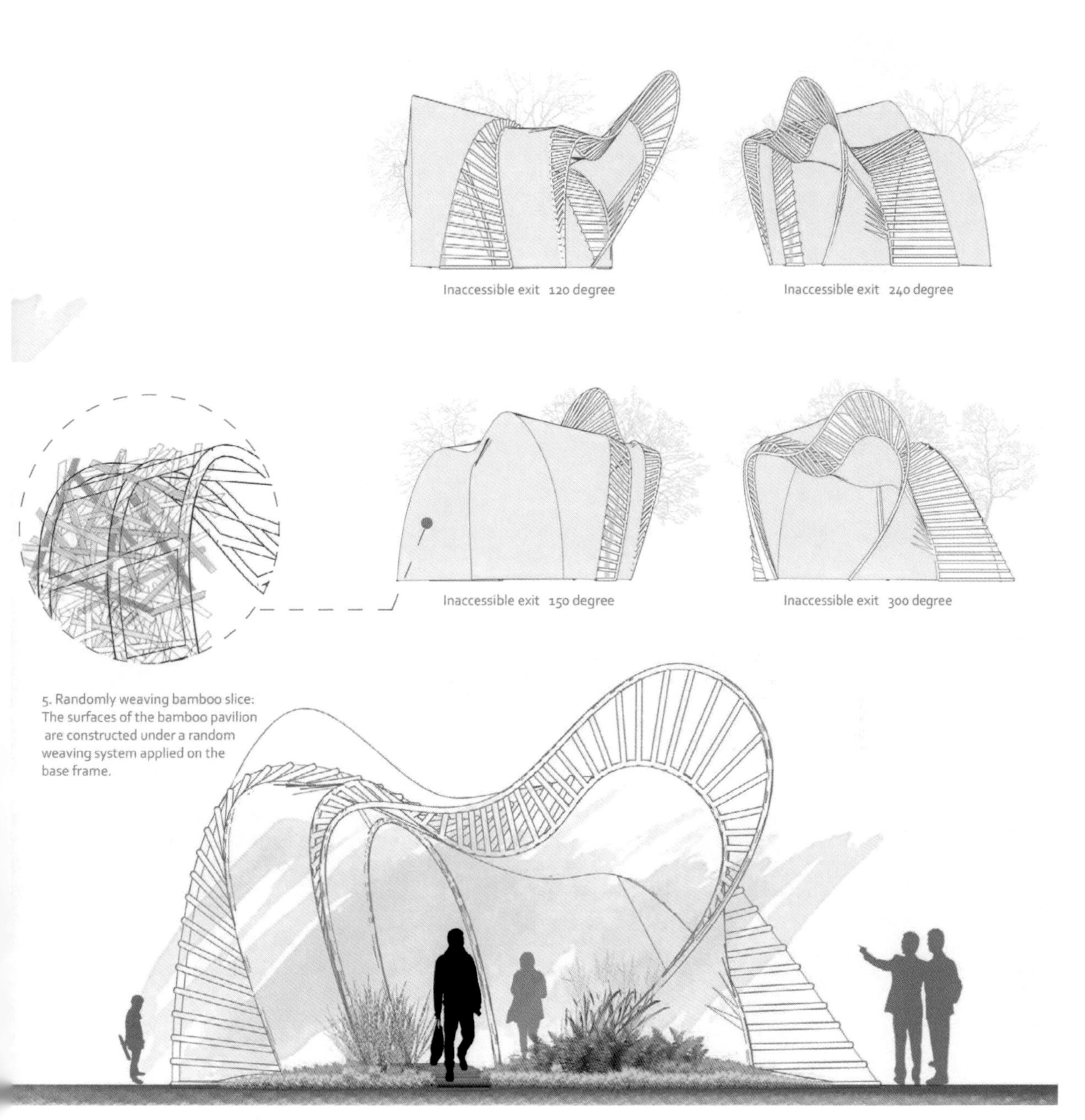
Inaccessible exit 120 degree
Inaccessible exit 240 degree
Inaccessible exit 150 degree
Inaccessible exit 300 degree
5. Randomly weaving bamboo slice:
The surfaces of the bamboo pavilion
are constructed under a random
weaving system applied on the
base frame.

子非鱼

ZI FEI YU

“子非鱼，安知鱼之乐？”
竹可弯，可直。
但曲直之辨，非吾本意。
尽态极妍，抑或曲折不定，更非竹之气节。
齐白石的一笔，又是出水之鳞。
谦虚，低调，是“鱼”，又非“鱼”。
一蓬草，一弯竹，
鱼，游，与乐。

简约的竹构、编织与组合，明晰的植物配置，纯粹的形体与空间，塑造朴素深远的景观意境，投射出平淡亲和的自然关照、含蓄内秀的品格与寓于意简于形的精神深度。

"You are not a fish, how can you know the enjoyment of fish? "(Zhuang Zi, 369 B.C.- 286 B.C.)
Bamboo, the very material, which is stubborn, while its flexibility is stunning. Bamboo can be bent or straight. However, our purpose doesn't simply distinguish the difference between bent or straight. Complex curves and dazzle appearance can't represent the inner spirits of bamboo.
It is not the integrity of a bamboo to be extremely beautiful, delicate or inflexible.
The brushstroke on Qi Baishi's painting is again a scale out of water. Modest, low-key, a fish, but not a fish.
Single curved bamboo, and a bunch of grass.
Leisurely, carefree and enjoyment.

This simple bamboo structure is to create pure form and space through weaving and combination as well as combined with clear plant configuration. Through shaping simple and profound landscape artistic conception, bamboo structure projects the insipid and friendly natural care, expresses the implicit and elegant character and the spiritual depth embodied in the simple meaning and shape.

参赛者学校：清华大学
指导老师：李树华、张学玲
参赛者：杨澈、魏欣辰、李静珂、王雅淇、江婷、杨临风、郭昊
School: Tsinghua University
Advisers: Li Shuhua, Zhang Xueling
Participants: Yang Che, Wei Xinchen, Li Jingke, Wang Yaqi, Jiang Ting, Yang Linfeng, Guo Hao

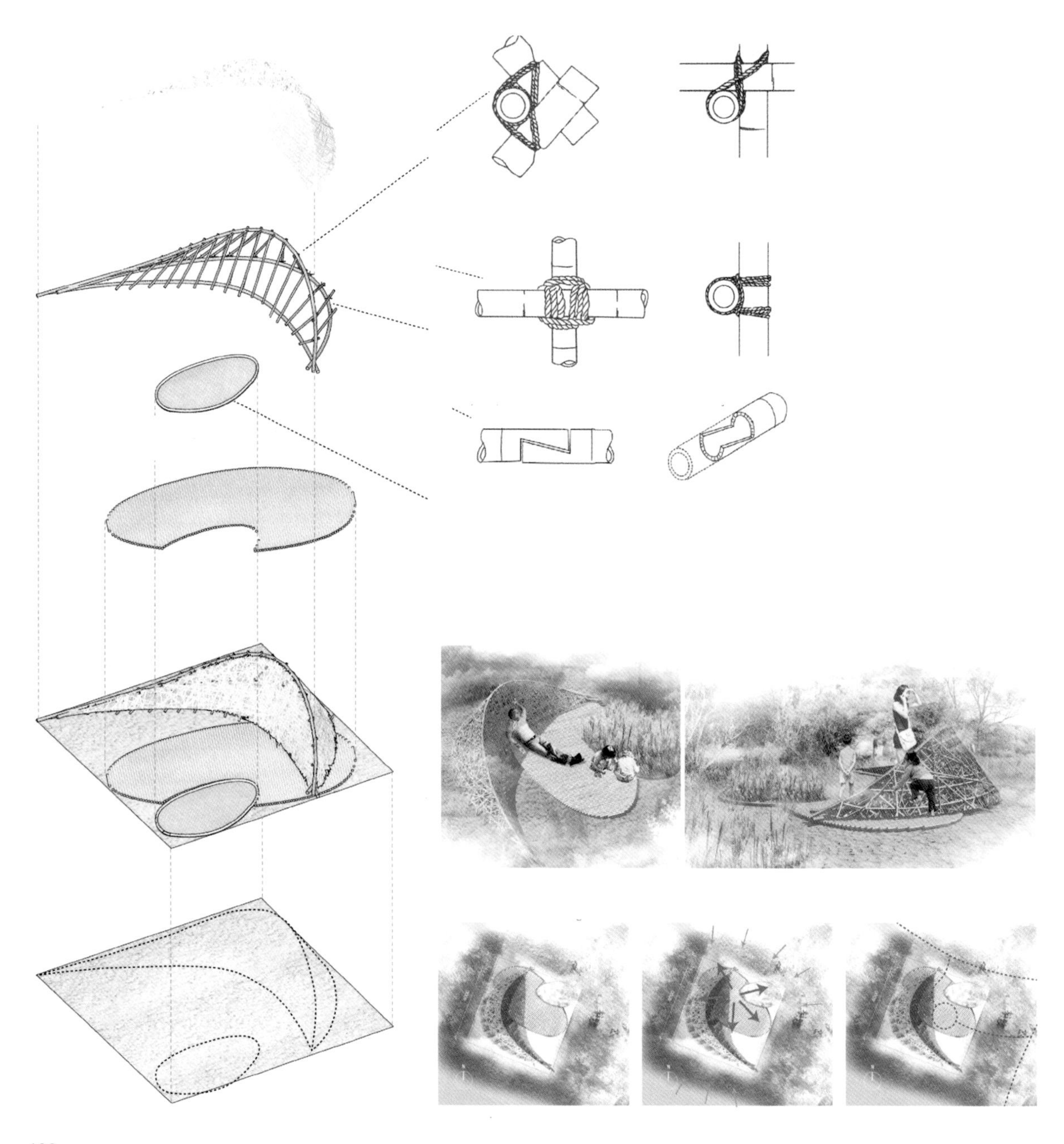

4000
4000
N

清荷风篁

LOTUS IN BREEZE

以“清荷风篁”为意象，以“小荷才露尖尖角”为展现方式，设计提取了荷花花瓣的元素，将四片花瓣围合成体验空间。弯曲扭转的竹凸显了花瓣的灵动感和流线空间，底板的竹采取同样竖向上的弯曲扭转形态，仿佛清风袭来，涟漪泛起，如“庭下如积水空明，水中藻荇交横，盖竹柏影也”的诗般意境。

With "a breeze with the fragrance of lotus in bamboo forest" as the imagery and "little lotus just shows its sharp points" as the display mode, the design extracts the elements of lotus petals and forms the experience space called bamboo forest with four petals. The twisted bamboo highlights the flexibility and streamline space of the petals, while the bamboo on the bottom adopts the same vertical twisting shape. The installation mimics the rippling atmosphere of a breeze as well as the poetic scene as "the moonlight shone through the courtyard and transparent as a pool of water. The water was crisscrossed with algae and weeds, which were the shadows of bamboo and cypress trees in the courtyard.

参赛者学校：重庆大学
指导老师：夏晖、罗丹
参赛者：邹宇航、冯晴、张立立、程语、丁昕昕、景巧琳、高艺飞、张宏海
School: Chongqing University
Advisers: Xia Hui, Luo Dan
Participants: Zou Yuhang, Feng Qing, Zhang Lili, Cheng Yu, Ding Xinxin, Jing Qiaolin, Gao Yifei, Zhang Honghai

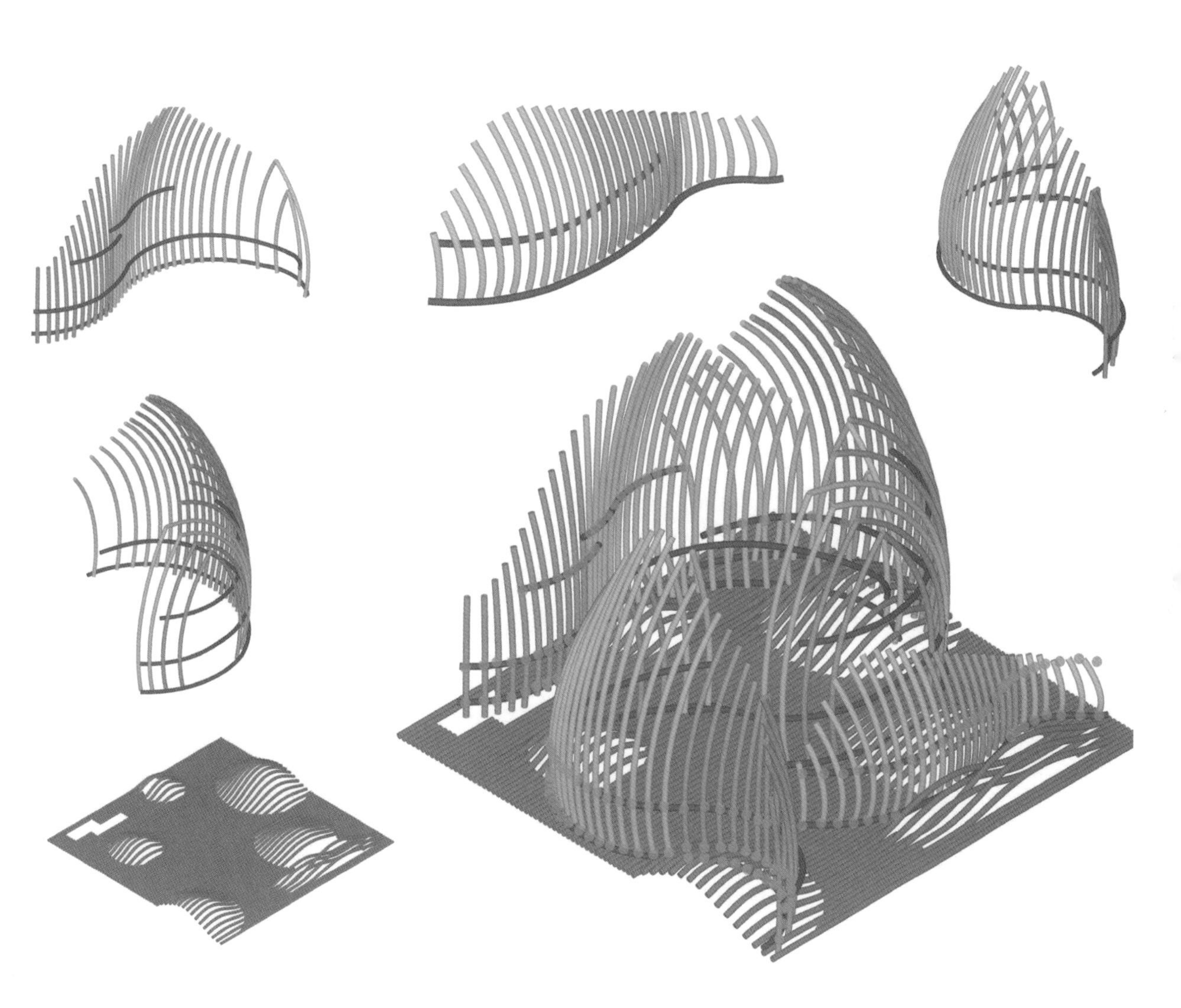

合叶栖
DWELLING IN LEAVES

本方案以“叶”这一种极小而莫大的形象为原型，用两片合上的叶子组合形成行进、转角、停留等模糊内外与功能的空间，并分别使用西侧冷色调蓝、紫色花境与东侧暖色调橙、黄色花境的色彩对比来暗示使用者在不同空间中安静与活泼的不同感受，将竹构与花园融合，从而展现对于现代城市人诗意花园的理解——喘息于现实生活的缝隙中，它是停驻点、休息处；用以容纳自己对生活的感悟，它是自我的感动理解。

Taking the minimal but maximal image of "leaf" as the prototype, This scheme uses two closed leaves to form a space with blurred interior and exterior functions such as marching, turning and staying. We also use the contrast color such as cold blue, purple border in the west and warm orange and yellow border in the east to indicate the different feelings of quiet and lively in different spaces. The project integrates bamboo structure with the garden so as to show our understanding of the poetic garden in modern cities which is breathing in the gap of real life. The bamboo structure is a stop and rest place to accommodate one's perception of life as well as a moving understanding of oneself.

参赛者学校：浙江农林大学
指导老师：洪泉、唐慧超
参赛者：杨憬铭、胡贝贝、郑朵拉、张雨卉、赵琪、肖达、黄胜孟
School: Zhejiang A & F University
Advisers: Hong Quan, Tang Huichao
Participants: Yang Jingming, Hu Beibei, Zheng Duola, Zhang Yuhui, Zhao Qi, Xiao Da, Huang Shengmeng

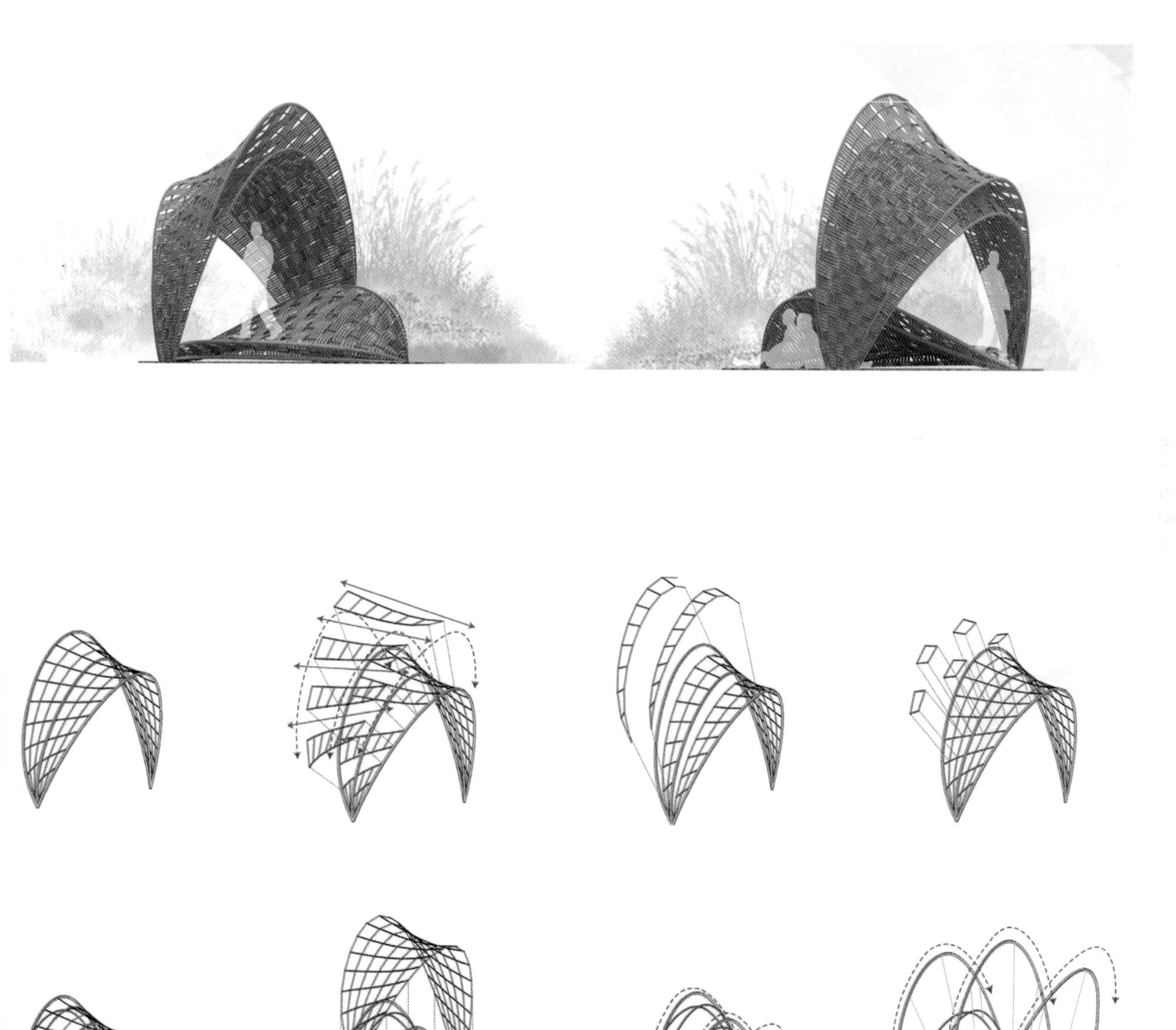

电光石火
VANISH IN A FLASH

“电光石火”比喻机锋敏捷，忽然触发，有所悟入。当今社会节奏快、压力大，碎片化的生活影响和塑造了青年的思维习惯，精力集中难度加大，注意力日益减弱，长文阅读能力下降。珍贵的休息时间也像闪电和石火一样转瞬即逝。

首先要利用好竹子轻盈灵巧、挺拔坚韧的特性，使设计打造出一种旋转、飞逝的动势。中部的弧形空间则会创造出一个供人休憩、畅谈、静坐之地，吸引人们来饮茶纳凉。同时竹片组合的表皮使得形态轻盈，与环境融合。

期望人们能充分利用破碎的休闲时光，进来感受竹子带来的一片清凉，让这难得的短暂更加美好，体会到自然带来的诗意！

"Vanishes in a Flash" means a metaphor for a sharp, suddenly triggered insight. In these days, society is fast-paced, stressful. Fragmented life influences and shapes the youngers' thinking habits, making it harder to concentrate or read a long text. Precious rest time also disappears like "vanishes in a flash".

We take advantage of bamboo's lightness, agility and tenacity, so that the design can create a rotating, flying momentum. The central arc space will create a place for people to have a rest, chat, sit quietly as well as attract people to drink tea. At the same time, the combination of bamboo skin makes the architectural form lighter and more integrated with the environment.

We hope that people can make good use of the rest time, make full use of their broken leisure time, come in to feel a piece of cool brought by bamboo. Just let this rare time more beautiful. Let everybody feel the poetry brought by nature!

参赛者学校：广州美术学院
指导老师：杨一丁、金涛
参赛者：马旭龙、龚博维、高万琪、李薇、郭超、潘泓、姜猛、陈桂玲
School: Guangzhou Academy of Fine Arts
Advisers: Yang Yiding, Jin Tao
Participants: Ma Xulong, Gong Bowei, Gao Wanqi, Li Wei, Guo Chao, Pan Hong, Jiang Meng, Chen Gulling

渡
FERRY

“春水碧于天，画船听雨眠”，满足了人们对江南的所有想象。一叶扁舟带你渡满江春水，载你渡一蓑烟雨，渡你寻满船清梦。轻舟八尺，低篷三扇，渡人、渡梦，也渡心。

该设计以江南乌篷船为原型，提取乌篷船中船篷的元素，将两只船篷进行叠加，意为在水中的两只船相遇而过。在地块西北角用竹片制成水波状，意为水上小舟。环绕构筑物主体的“江心洲”内部搭配种类层次丰富的盆栽花卉，创造丰富多彩的视觉体验。此外，在构筑物物件上也运用了竹作为原材料，船篷前悬挂用竹子制作的风铃，夜晚“江心洲”内部放置镂空竹灯，营造出一个雅致的冥想空间。

"The river in spring is clearer and greener than the sky is blue. People can also fall asleep on painted boats listening to the rain." This ancient poem satisfies all people's imaginations of regions south of Yangtze River. A flat boat takes you to cross the river spring water, drive you to cross the misty rain, accompany you to find full of boat clear dream. The boat is light with three low canopies to guide the people, help the dream and save the heart.

The design takes caravan as the prototype. It extracts the elements of canopies in caravan and overlays two canopies, which means that two boats in the water meet. In the northwest corner of the plot, bamboo pieces are made into water waves, meaning boats on the water. Around the main body of the structure "Jiangxinzhou" inside, the variety and layers of potted flowers are matched out to create a rich and colorful visual experience. In addition, bamboo is also used as the raw material for structures and objects. Wind chimes made of bamboo hang in front of the ship's awning. At night, hollow-out bamboo lamps are placed inside the "Jiangxinzhou" to create an elegant meditation space.

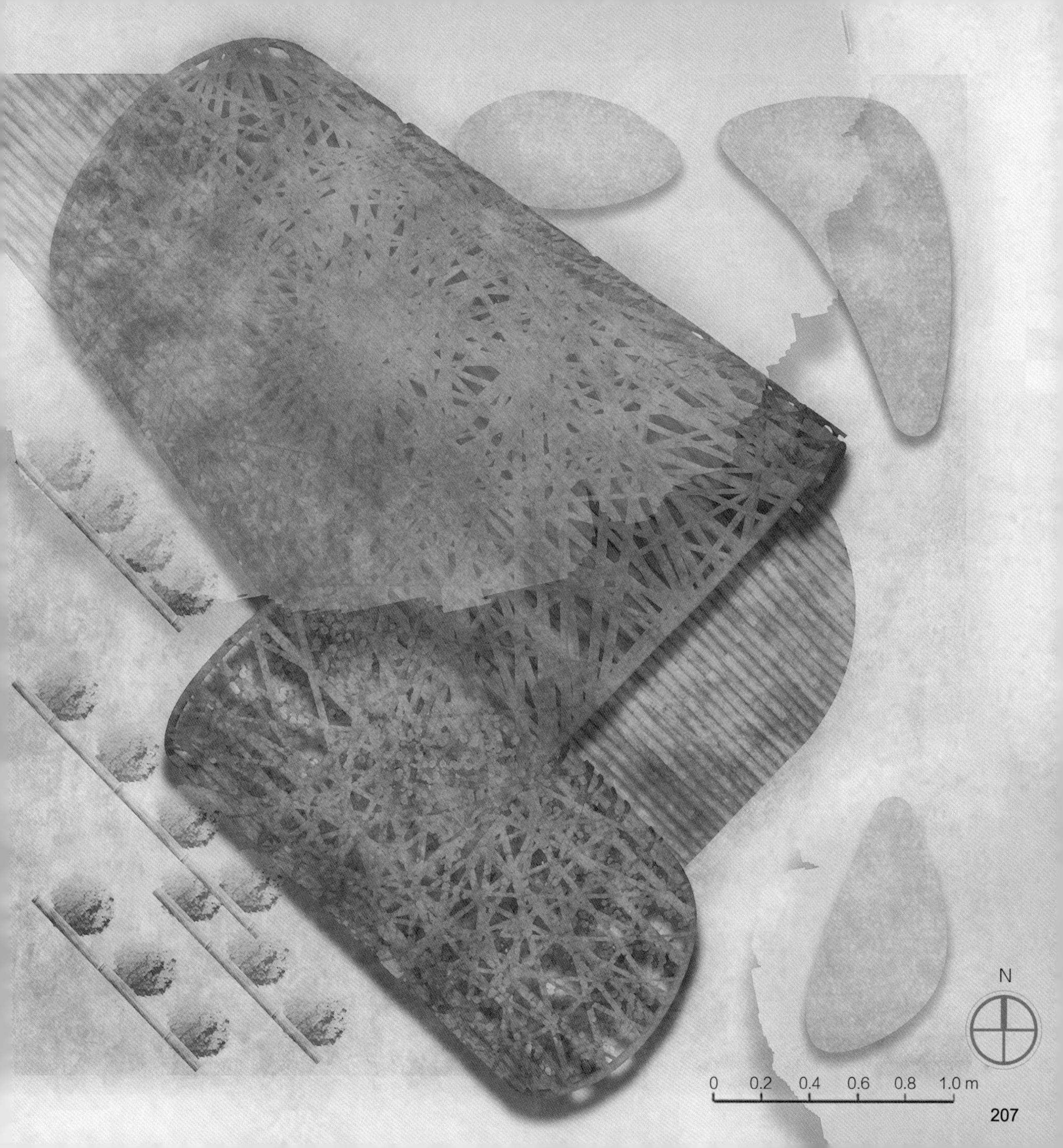
N
0
0.2
0.4
0.6
0.8
1.0 m

卧云端
LYING ON CLOUDS

方案灵感来源于“云”和“卧”的概念，加之对诗意栖居方式的思考凝练，利用竹结构较高的抗拉和抗压强度，对空间进行有机组合，形成“卧云端”的整体构思。方案通过高、中、低三个层次的空间变化和坐、倚、卧的使用功能展示一种闲适的状态，在高强度、快节奏的工作中余留一方天地，放松身心，感受生活。

This scheme is inspired by the shape of "cloud" and the concept of "lying", coupled with the thought of a poetic dwelling way. Bamboo has a high tensile and compressive strength and contributes to an organic and poetic space combination, which forms the overall idea of "lying on clouds". Its three levels of spatial structure (high, middle and low) and different functions (for sitting, laying and relying) show a poetic and leisure state, remaining a little free world in the high-intensity and fast-paced study/work life for us to return to nature, relax our body and feel life.

参赛者学校：西北农林科技大学
指导老师：张新果、杨梅花
参赛者：孟凡琦、刘浩琳、冯懿梦、陈进、洪浩庭、王逸涵、洪阳、于重光
School: Northwest A & F University
Advisers: Zhang Xinguo, Yang Meihua
Participants: Meng Fanqi, Liu Haolin, Feng Yimeng, Chen Jin, Hong Haoting, Wang Yihan, Hong Yang, Yu Chongguang

风巢

THE WIND NEST

风，以动御万物；巢，博众家之长。本设计灵感正是来源于风御巢动，群星相聚，以轻盈的竹构做灵感之巢，以摇曳的花境展星聚之势。“风巢”指学生们创作的灵感如同夜空中璀璨的星星，在激荡间碰撞出火花。“风巢”亦可读为“风潮”，“筑巢引风”，表现了风华正茂的青年学生创造新时代的魄力。

在空间设计上，穿过竹构的小径将花境和竹境连为一体，营造“人在竹境内，犹在轻风中”的意境。结构设计上，通过搭建实景模型，既验证了方案的可实施性，也可充分展现竹子轻盈、柔韧的竹性。花境设计上，地面流线造型模仿风的轨迹，实现风和竹构的交融。同时，选用芒草作为主要植被材料，营造轻松、惬意、如风一般、随风飘动的诗意境界。

Wind uses the momentum to rein all. Nest gathers the strength of everyone. This design's inspiration comes from the the nest in the wind. It used the lighter bamboo structure to create a nest full of inspiration, while swaying flower border was to show the nest that gathering the stars. "Wind Nest" refers that the inspiration of students is like the shining stars, which collided sparks in the agitation. "Wind Nest" can also be considered as "Trends" that can attract the wind, which demonstrates the courage of young students to create a new era.

In the space design, the path through the bamboo structure connects the flower border and the bamboo structure into a unified whole, decreasing the distance among visitors, vegetation and bamboo structure as well as creating the experience of "people walk in the bamboo structure, just like in the light wind". As for the structural design, by building a real model, we not only verify the feasibility of the scheme, but also fully demonstrate the lightness and flexibility of bamboo. In the flower border design, the streamlined shape on the ground mimics the trajectory of the wind, realizes the mutual interweaving of the wind and the bamboo structure. At the same time, we use multiple miscanthus as vegetation materials, so that it can create a poetic realm that is relaxed, cozy, windy, and fluttering in the wind.

参赛者学校：天津大学
指导老师：王洪成、胡一可
参赛者：陈丽君、邓剑、孙雅伟、张尔科、郭茹、刘润潼、张浩、张佳乐
School: Tianjin University
Advisers: Wang Hongcheng, Hu Yike
Participants: Chen Lijun, Deng Jian, Sun Yawei, Zhang Erke, Guo Ru, Liu Runtong, Zhang Hao, Zhang Jiale

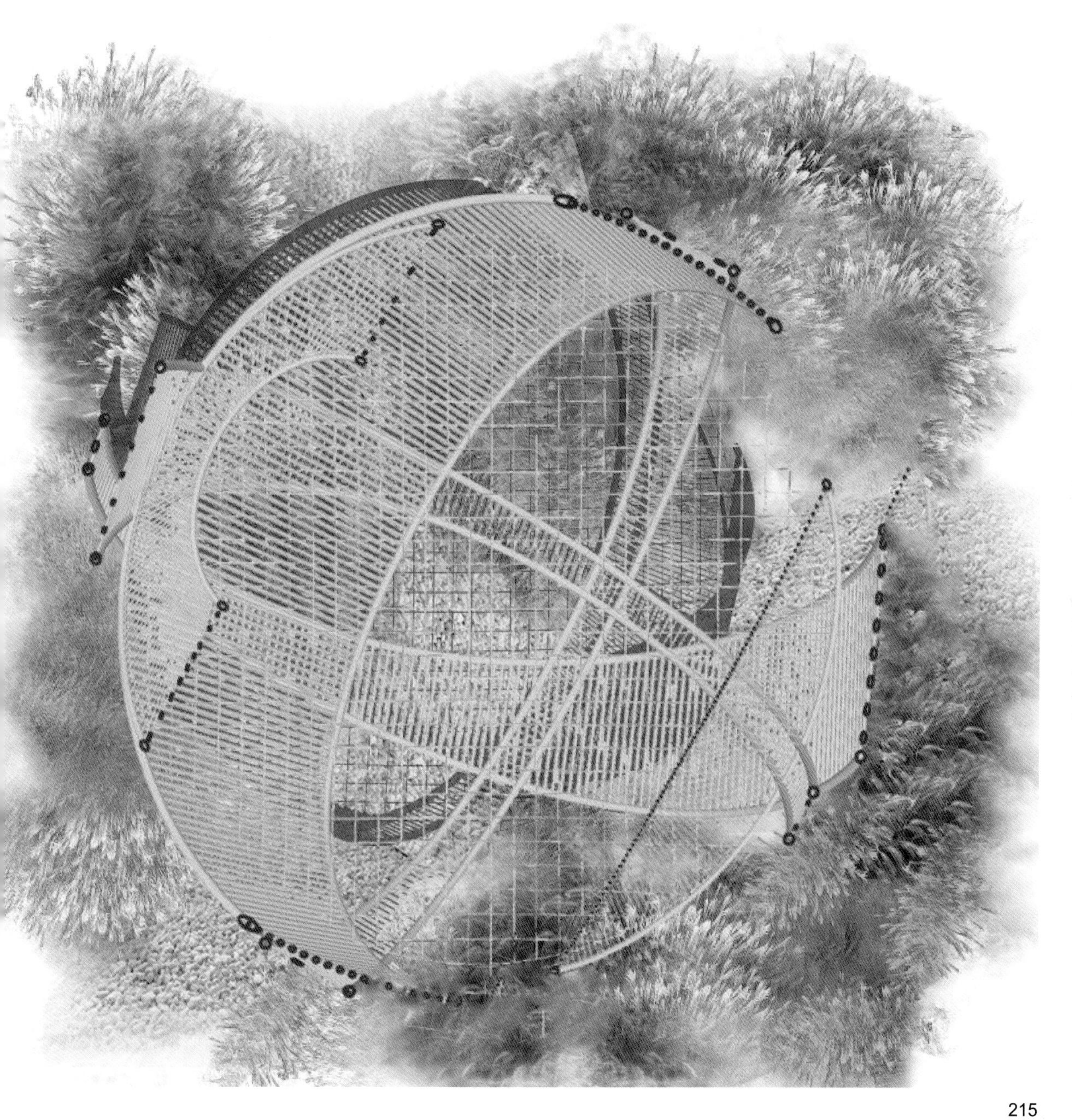

第三届（2020）北林国际花园建造节

The 3rd (2020) BFU International Garden-Making Festival

竞赛主题：
秘境花园
建造主题：
公园城市，花重锦城——成都生活美学，天府花园魅力

方案征集时间：2020 年 4-9 月
实地建造时间：2020 年 10 月 20-23 日
开放活动时间：2020 年 10 月 23 日 -11 月 24 日
举办地点：成都市桂溪生态公园

报名选手：
241 组设计团队，来自 77 所高校，共 1471 人
实地建造团队：
入围团队 17 个、受邀参加团队 8 个

指导单位：
中国风景园林学会、国际竹藤组织
主办单位：
中国建筑学会园林景观分会、成都市商务局、成都市体育局、成都市新闻出版局
承办单位：
成都市风景园林学会、北京《风景园林》杂志社有限公司、成都天府绿道文化旅游发展股份有限公司、成都天府绿道建设投资有限公司、成都市公园城市建设发展研究院、成都市国际商务会展服务中心、成都市花木技术服务中心
支持单位：
中国风景园林学会教育工作委员会、成都市公园城市建设管理局、北京林业大学园林学院、成都市博览局
协办单位：
成都市公园城市园林绿化事业发展中心、成都市绿化工程队、成都市绿化工程三队、成都市林草种苗站、成都市草堂花圃、成都大熊猫繁育研究基地、成都动物园、成都市人民公园、成都市植物园、成都市望江楼公园、成都市百花潭公园、成都市文化公园、成都市园林建设技术服务中心、成都市湿地保护中心

Competition Theme:
The Garden of Mystery
Construction Theme:
Park City, Blossom City — Garden's charm presenting aesthetics of civic life in Chengdu

Call for Entry: April–September, 2020
On-site Construction: October 20–23, 2020
Opening: October 23–November 24, 2020
Location: Chengdu Guixi Ecological Park

Registered Teams:
241 teams, from 77 universities, a total of 1,471 students
Construction teams:
17 finalist teams, 8 invited teams

Guidance Organizers:
Chinese Society of Landscape Architecture, International Bamboo and Rattan Organization
Organizers:
Institute of Landscape Architecture of the Architectural Society of China, Chengdu Bureau of Commerce, Chengdu Bureau of Sports, Chengdu Bureau of Press and Publication,
Executive Organizers:
Chengdu Society of Landscape Architecture, Beijing Landscape Architecture Journal Ltd, Chengdu Tianfu Greenway Culture and Tourism Development Co., Ltd., Chengdu Tianfu Greenway Construction Investment Co., Ltd., Chengdu Park City Construction and Development Research Institute, Chengdu International Business and Exhibition Service Centre, Chengdu Plantation Technical Service Center
Support Organizers:
Education Committee of Chinese Society of Landscape Architecture, Chengdu Municipal Park City Construction and Management Bureau, School of Landscape Architecture of Beijing Forestry University, Chengdu Municipal Bureau of Exposition
Co-organizers:
Chengdu Park City Landscape Development Center, Chengdu Greening Construction Team, Chengdu 3rd Greening Construction Team, Chengdu Grass Seedlings Station, Chengdu Thatched Cottage Flower Cultivation Garden, Chengdu Research Base of Panda Breeding, Chengdu Zoo, Chengdu People's Park, Chengdu Botanical Garden, Chengdu Wangjianglou Park, Chengdu Baihuatan Park, Chengdu Cultural Park, Chengdu Garden Construction Technical Service Center, Chengdu Wetland Protection Center

庄生梦·迷蝶
IN DREAM WITH THE BUTTERFLY

庄生梦·迷蝶是轻盈的，梦幻的。
正如《庄子·齐物论》中庄周的梦蝶之境，在路径穿梭中达到“物我两相忘”。

空间回环而曲折，密径通幽，芒草掩映。蝴蝶翅膀的优雅弧度与竹篾的编织生动结合，双层圆圈构建成蝶翼的表皮纹理。在有限的空间中尽可能丰富视觉感受与空间形态，打破内外空间的分割对立，使造型舒展起伏，无分彼此，亦梦亦真。

在花园的秘境探索中回溯自然与童真，寻找自己内心的秘境。

In Dream with the Butterfly is light and dreamy, just like Zhuang Zhou's dream of butterfly in "on Leveling All Things", achieving the goal of forgetting everything in the shuttle of paths.

The space is circuitous and tortuous, with winding paths leading to tufted awn grass. The graceful radian of butterfly wings is combined with the weaving of bamboo strips, and the surface texture of butterfly wings is constructed by double-layer circles. In the limited space, this scheme try its best to enrich the visual experience and spatial form, break the division and conflict between the internal and external space. The shape stretches up and down. It is not only a dream but also a reality.

In the exploration of the mysterious garden, people can trace back to nature and innocence, finding their own inner secret.

参赛者学校：北京林业大学
指导老师：郑小东、张诗阳
参赛者：廖丹妍、杨艺文、郑睿楠、谭铃千、赵祎祺、吴媛玉
School: Beijing Forestry University
Advisers: Zheng Xiaodong, Zhang Shiyang
Participants: Liao Danyan, Yang Yiwen, Zheng Ruinan, Tan Lingqian, Zhao Yiqi, Wu Yuanyu

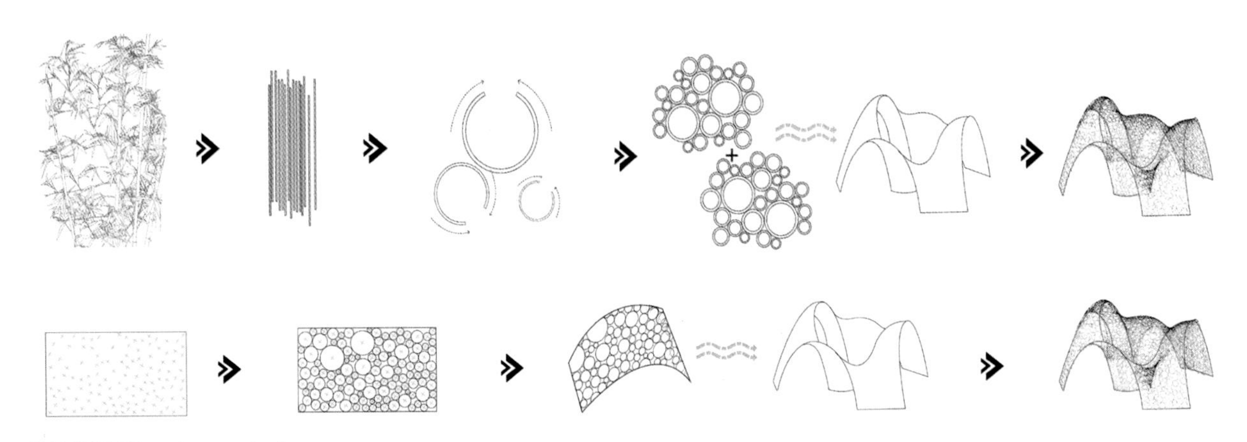

编织分析 /Weaving analysis

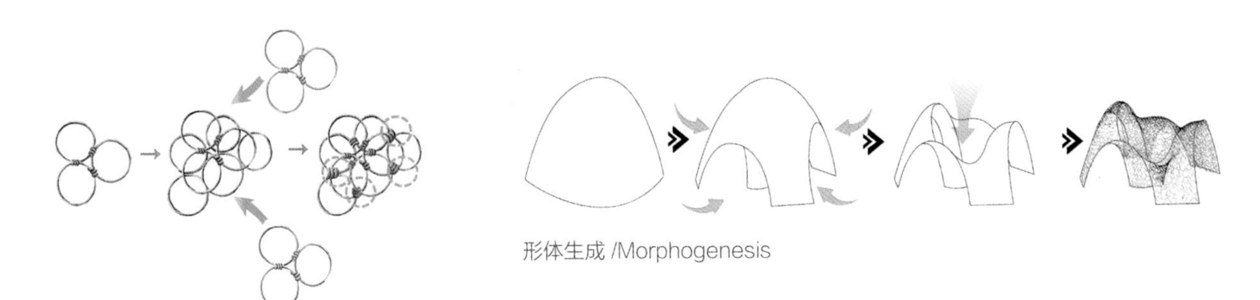

形体生成 /Morphogenesis

编织细节 /Weaving detalis

立面图 /Elevations

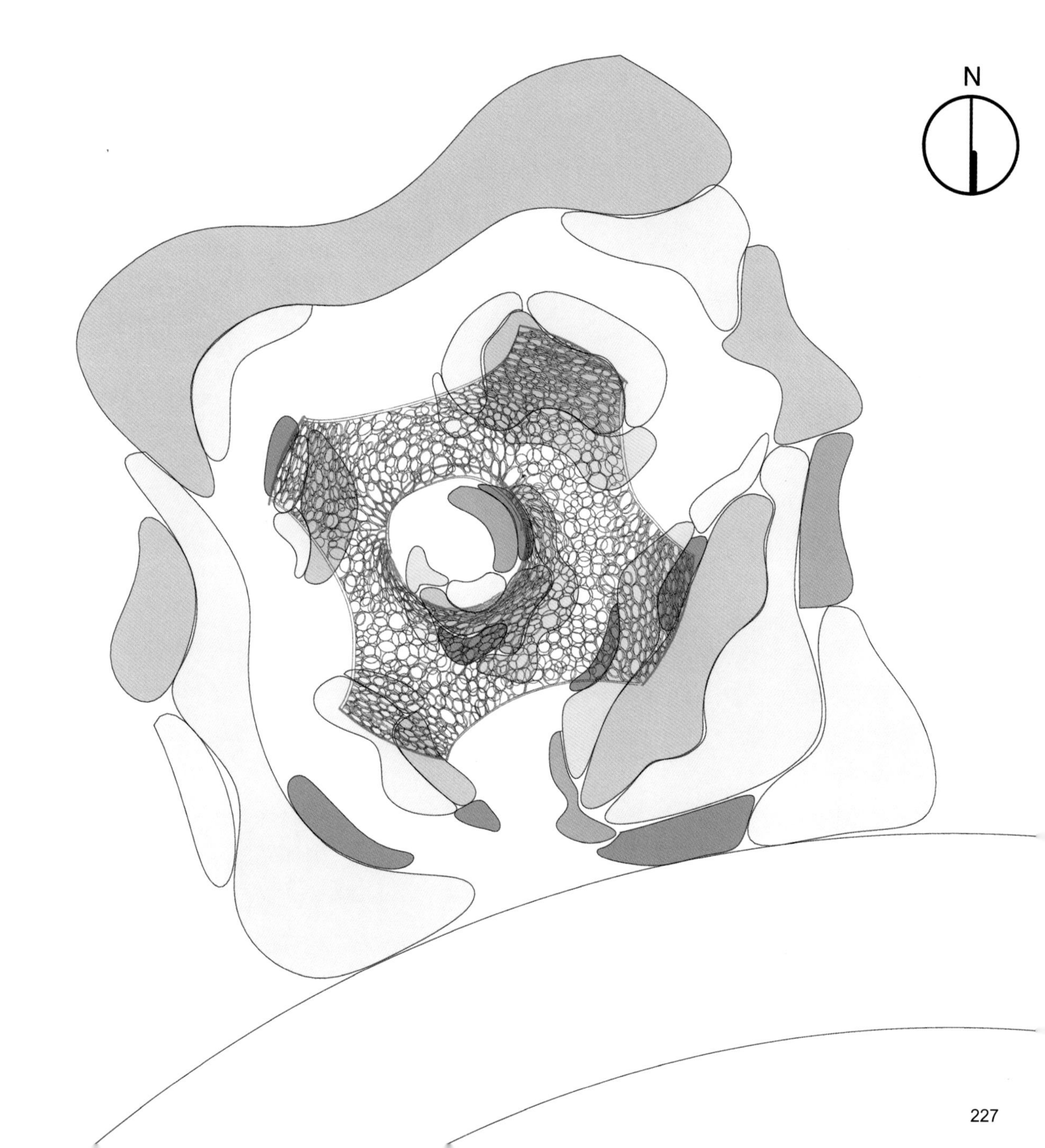
N

帘幕卷
PICTURE SCROLL OF BAMBOO CURTAIN

“画栋朝飞南浦云，珠帘暮卷西山雨”。方案根据“珠帘幕卷”的理念展开设计，希望花竹相融，帘幕载卷。竹片交织形成竹帘，竹帘旋转生成帘幕，帘幕叠加构成画卷。力争在符合力学逻辑与美学原理基础上，使构筑物的结构、表皮、空间、形态一体化。以纯粹的编织形式建造竹构，由简入繁，亦实亦虚，境融竹中，形成亦梦亦幻的“帘幕卷”秘境花园。

花园意境营造与竹构干净、整洁的形式相呼应。花境设计采用曲线形式，利用多叶植物烘托意境，同时用观赏草进一步围合空间，营造神秘的空间体验。花园底部用卵石和碎石铺地，结合弯曲的圆竹形成自然的路径，与竹构相统一，营造了一座吸引人进入、探索的秘境花园。

The project is designed according to the concept of "bead curtain roll", hoping that the flowers and bamboo will blend together and the curtain will be rolled up. The bamboo strips are interwoven to form a bamboo surface, which rotates to create a curtain, and the curtains are superimposed to form a picture scroll. The project strives to integrate the structure, skin, space, and form on the basis of conforming to the mechanics logic and aesthetic principles. By constructing the bamboo structure with a pure weaving form, which is both real and virtual and starts from simplicity to complexity, merging the environment with the bamboo, this project forms a "Picture Scroll of Bamboo Curtain" secret garden that is also dreamy and illusory.

The artistic conception of the garden echoes the clean and tidy form of the bamboo structure. The flower border design adopts curve form, and uses leafy plants to build an artistic conception. At the same time, the space was further enclosed by ornamental grasses, creating a mysterious space experience. The ground of the garden is paved with pebbles and gravel, combined with curved round bamboos to form a natural path, which is unified with the bamboo structure, creating a secret garden that attracts people to enter and explore.

参赛者学校：天津大学
指导老师：王洪成、胡一可
参赛者：郭茹、张文正、邓唐敏、顾阳、臧青茹、温雯、杨婷婷、赵玥
School: Tianjin University
Advisers: Wang Hongcheng, Hu Yike
Participants: Guo Ru, Zhang Wenzheng, Deng Tangmin, Gu Yang, Zang Qingru, Wen Wen, Yang Tingting, Zhao Yue

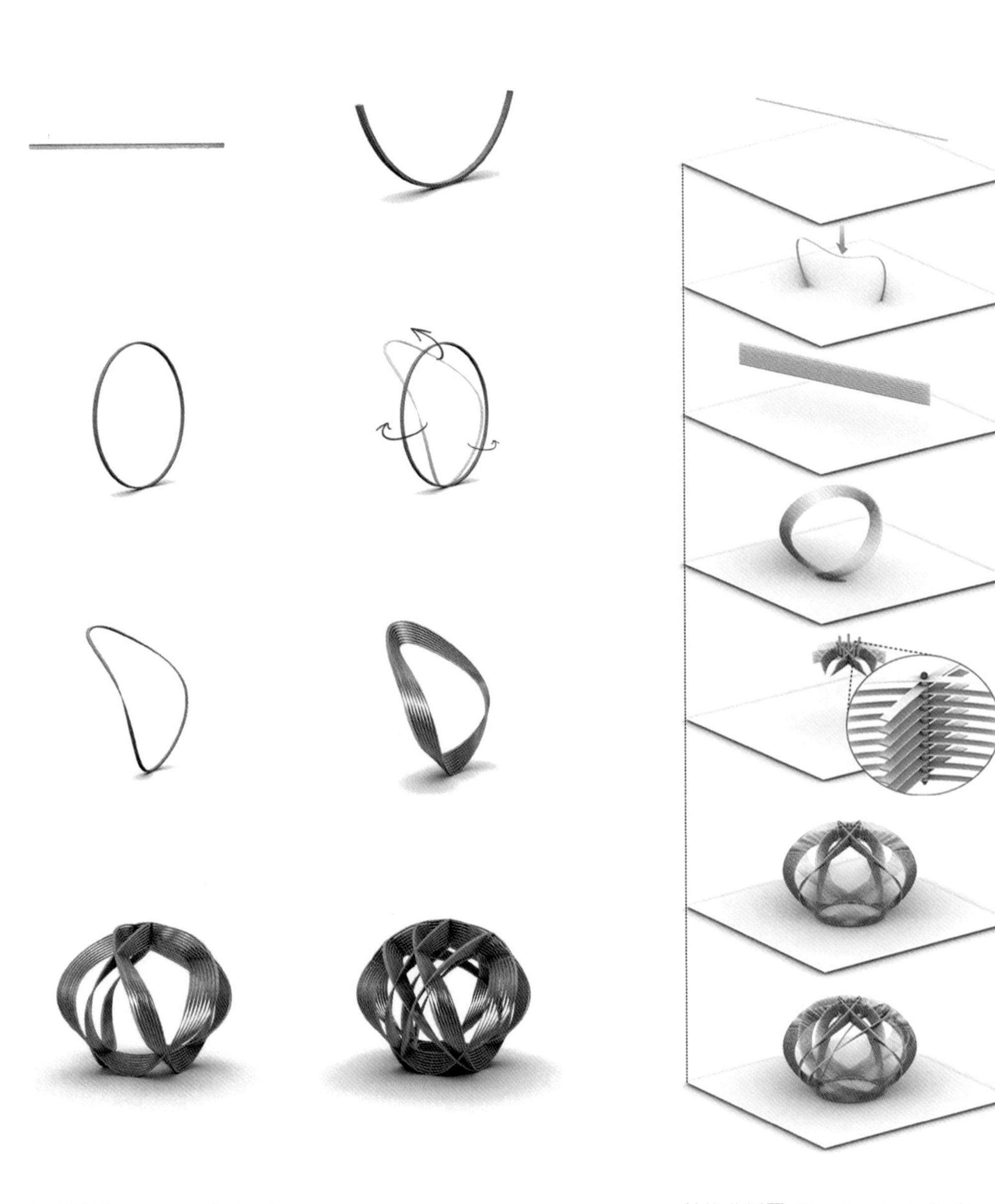

形态推演图 /Form deduction

结构分析图 /Structure analysis

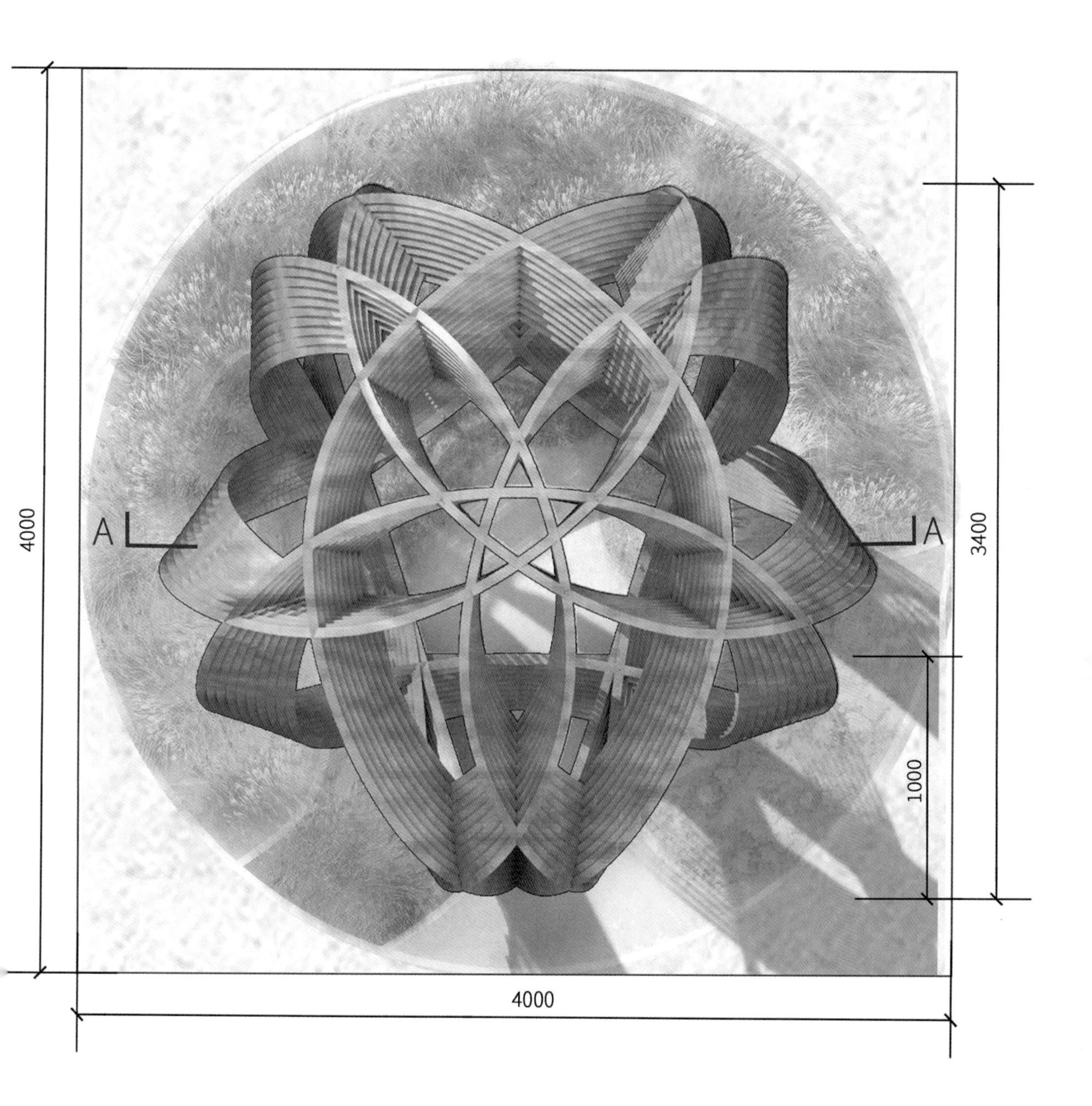
4000
A
A
3400
1000
4000

14号地块
帘幕卷
Picture Scroll
Of Bamboo Curtain
参赛院校
天津大学
指导老师
参赛学生

森之秘语

THE WHISPER OF FOREST

在城市中面对生活压力和工作烦恼的设计师兔子先生，每天辛苦地设计、画图、改方案，他忘了自己存在的意义，在城市中逐渐活成了一个孤岛，生活不再快乐，设计也没有了灵感……

直到这一天，他在飞快行驶的地铁上发现窗外的城市高楼逐渐变成了昏暗的森林，地铁消失了。他在昏暗的树林里开始打盹，梦中出现了一个神秘且富有安全感的树洞。醒来之后他发现自己回到了充满阳光的秘境森林，在一片苇草中他发现了那个正在搭建的树洞。他与这位搭建树洞的住在森林里的快乐的兔子相对而坐，倾吐心声，释放压抑在心里很久的种种烦恼。

兔子先生走出树洞，他才发现那只快乐的兔子就是曾经的自己，兔子先生终于找回了迷失的自我，发现眼前另一片美丽的花园……

Mr. Rabbit, a designer who faces the pressure of life and worries about his work in the city, works hard every day to design, draw and change plans. He forgets the significance of his existence and gradually becomes a lonely island in the city. Life is no longer happy, and there is no inspiration for design.

Until this day, he found that the high-rise city buildings outside the subway window gradually turned into a dark forest, and the subway disappeared. He began to take a nap in the dark woods. In his dream, a mysterious and safe tree hole appeared. After the nap, he found himself back to the forest full of sunshine. In a reed grass, he found a tree hole being built. He sat opposite the happy rabbit who built a tree hole and lived in the forest. He poured out his heart and released all kinds of worries that had been suppressed in his heart for a long time.

When Mr. Rabbit walked out of the tree hole, he found that the happy rabbit was who he used to be. Mr. Rabbit finally regain his lost self and found another beautiful garden in front of him.

参赛者学校：重庆大学
指导老师：夏晖、罗丹
参赛者：汤思琦、廖文静、叶馨、李丝倩、高群、黄荧、冯诗雁、尹子佩
School: Chongqing University
Advisers: Xia Hui, Luo Dan
Participants: Tang Siqi, Liao Wenjing, Ye Xin, Li Siqian, Gao Qun, Huang Ying, Feng Shiyan, Ying Zipei

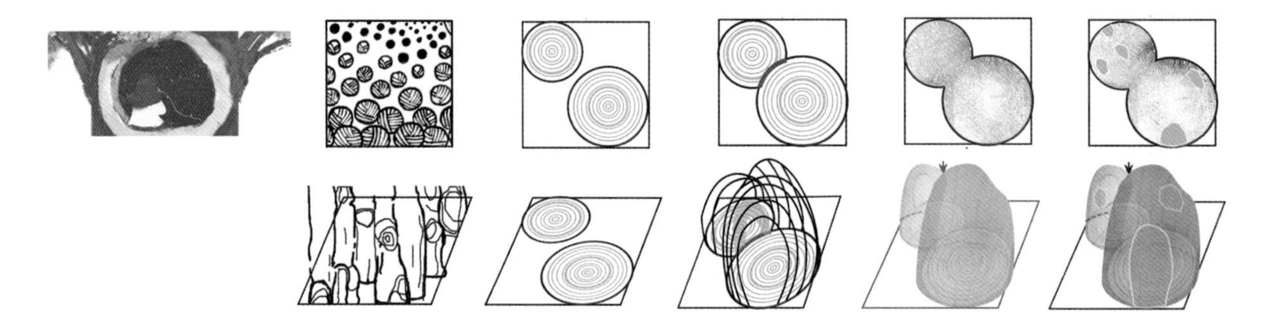

概念 /Concept

地面骨架 /Ground skeleton

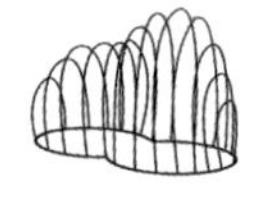

竖向骨架 /Vertical framework

横向骨架 /Transverse framework

表面开洞 /Surface hole

内层编织 /Inner braiding

外层编织 /Outer weave

结构 /Structure

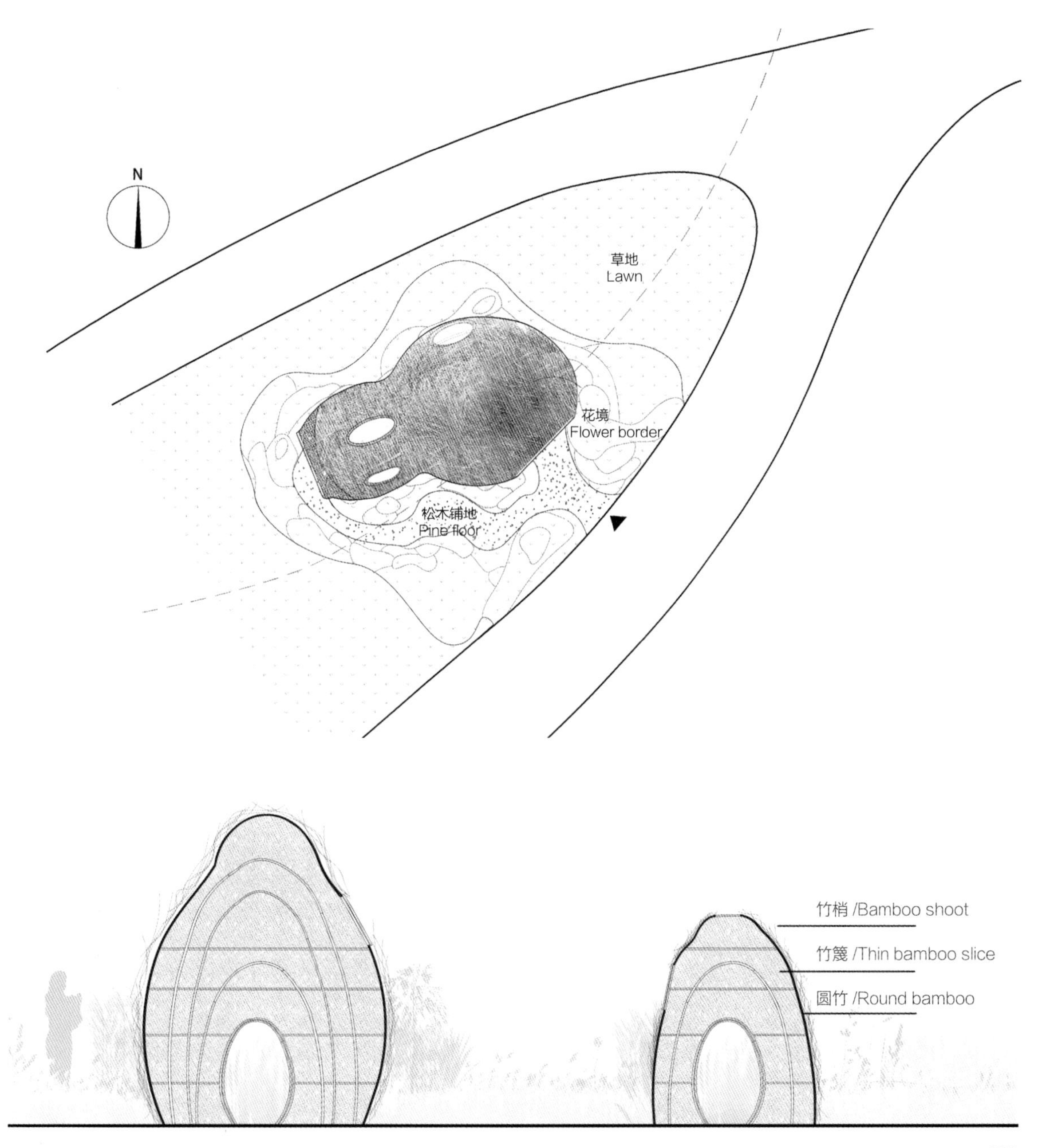
N
草地
Lawn
花境
Flower border
松木铺地
Pine floor
竹梢 /Bamboo shoot
竹篾 /Thin bamboo slice
圆竹 /Round bamboo

迷笙
BAMBOO ORGAN

什么是秘境?

我们从王维的《竹里馆》中找到了答案——

幽深竹林里弹琴长啸，独坐辋川时明月高悬。

虽竹影掩映，琴声音却透叶而来，秘境中“乐”的存在不可或缺，人在其中，视线受阻，听到的却绵延。

根据上述对秘境的设想，我们提取竹林、竹笋、竹根三个元素，结合中国传统乐器——笙去塑造视听结合的观感体验，创造了一个独自一方天地的花园、一处朦胧而幽邃的秘境。

What is the mysterious world?

We find the answer in Wang Wei's "Bamboo Pavilion":

Deep in the bamboo forest, playing the organ and singing, sitting alone in Wangchuan, while the moon is hanging high.

Although the bamboo shadow is set off, the music comes through the leaves, the existence of "music" in the secret world is indispensable. The sight of people inside is blocked, but the hearing is continuous.

Based on the above assumption of the mysterious world, we extract three elements of bamboo forest, bamboo shoot and bamboo root, and combine them with the traditional Chinese musical instrument “sheng” to create a visual and visual experience, forming a garden alone in heaven and earth, a hazy and secluded place.

参赛者学校：北京林业大学、清华大学
指导老师：郑小东、段威
参赛者：高嘉阳、赵书涵、孙千翔、朱临昕、马迎雪、韦明洁、张佳艺、乌家宁
School: Beijing Forestry University, Tsinghua University
Advisers: Zheng Xiaodong, Duan Wei
Participants: Gao Jiayang, Zhao Shuhan, Sun Qianxiang, Zhu Linxin, Ma Yingxue, Wei Mingjie, Zhang Jiayi, Wu Jianing

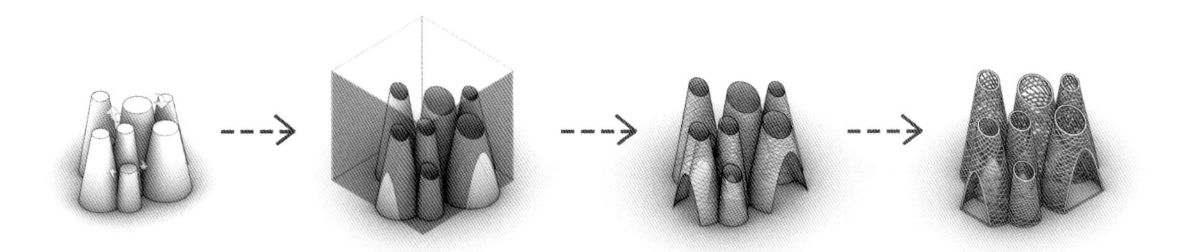

形态生成 /Formation

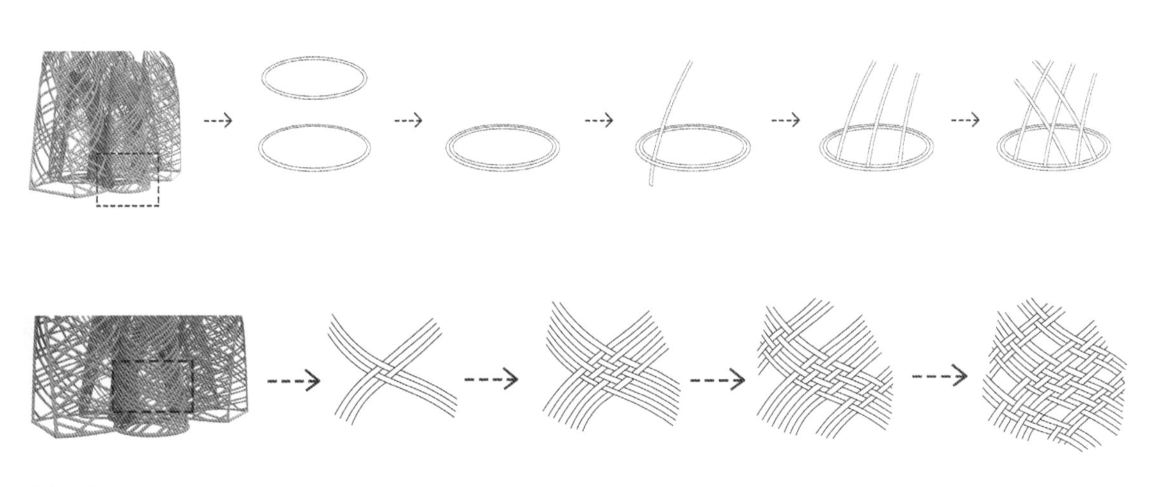

力糅编织 /Weaving

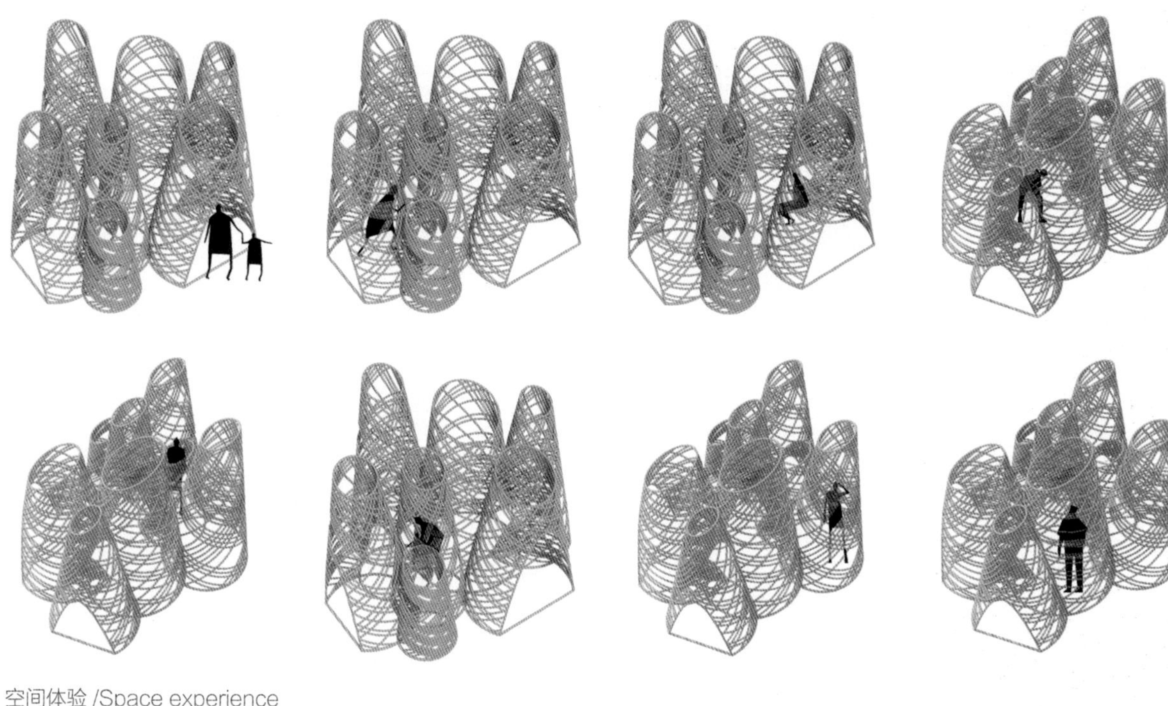

空间体验 /Space experience

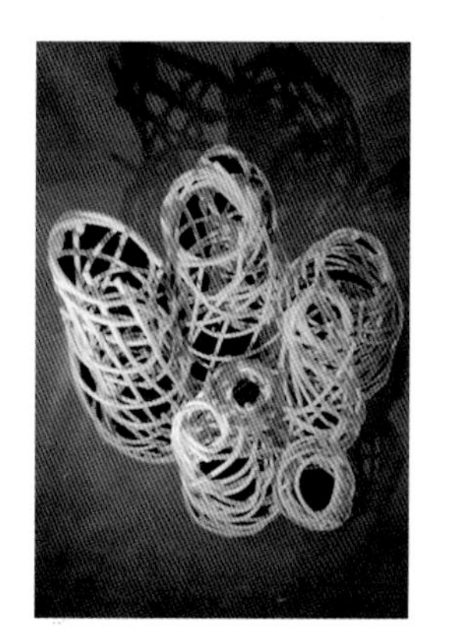

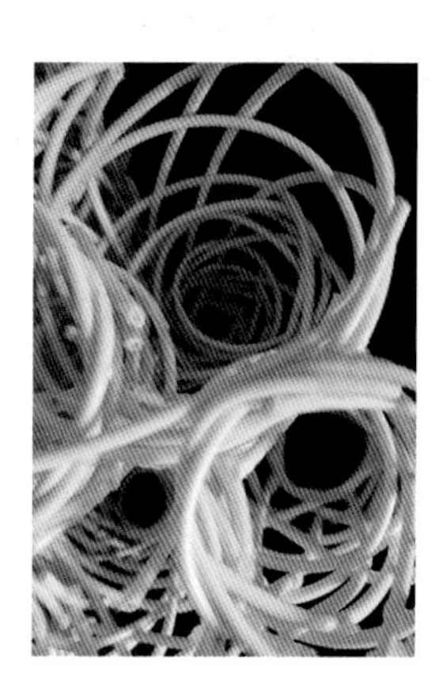

手工模型 /Manual model

细胞魔方
MAGIC CELL CUBE

秘境除肉眼可见的世界外，还有深藏于微观世界的自然神秘。本方案将细胞的生长过程变化抽象拆解为一个个魔方小块空间，利用竹篾柔软特性打造空间的纵深感与螺旋感，展现不同的生命变化；通过相似植物的色彩、高度变化象征着生命的逐步成长过程。最后各部分相互组合形成了一个内部变化莫测、外部整合统一的神秘空间——细胞魔方。

In addition to the visible world, there are hidden mysteries in the microscopic world of nature. This scheme abstracts the changes in the growth process of cells into a rubik's cube space, and uses the soft characteristics of bamboo strips to create a sense of depth and spirality in the space, showing different life changes.The gradual growth of life is symbolized by the color and height changes of similar plants. Finally, the various parts combine to form a mysterious space with internal changes and external integration—the Magic Cell Cube.

参赛者学校：华南农业大学
指导老师：李梦然、李健
方案阶段参与者：张译雯、刘康、杨楠、黄冰怡、陈漫婷、饶贵珊、苏法文
建造阶段参与者：杨楠、张译雯、刘康、刘蓉、黄曼仪、陈沐华、杨晴
School: South China Agriculture University
Advisers: Li Mengran, Li Jian
Conceptual Design Participants: Zhang Yiwen, Liu Kang, Yang Nan, Huang Bingyi, Chen Manting, Rao Guishan, Su Fawen
Construction Participants: Yang Nan, Zhang Yiwen, Liu Kang, Liu Rong, Huang Manyi, Chen Muhua, Yang Qing

概念生成
Concept Generation

细胞：生命单元 → 竹子：生命的张力 → 魔方：多变和整合

Cell: life unit → Bamboo: the tension of life → Magic cuber: variety and integration

孕育
Born

竹篾螺旋编织
Bamboo strips are spirally woven

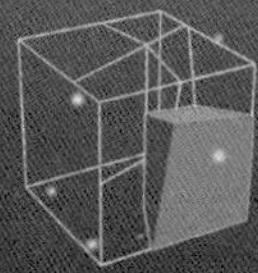

入口空间—神秘
Entrance space – mystery

逐步成长
Grow

竹篾向心排列
The bamboo strips are aligned to centrl

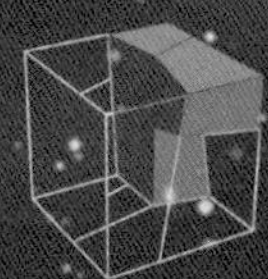

成长空间—密闭
Growth space – closed

细胞魔方
Magic Cell Cube

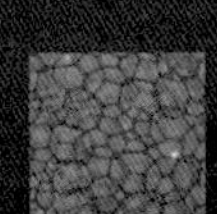

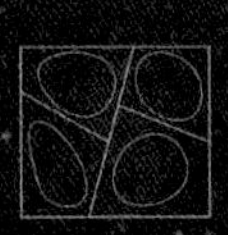

迸发
Burst out

减少排列密度
Reduce arrangement density

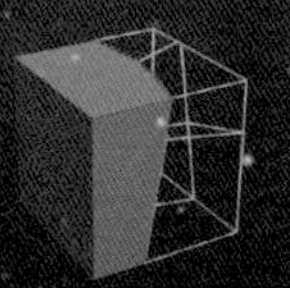

迸发空间—开阔
Burst space – openness

无序到有序的整合
disorder to order

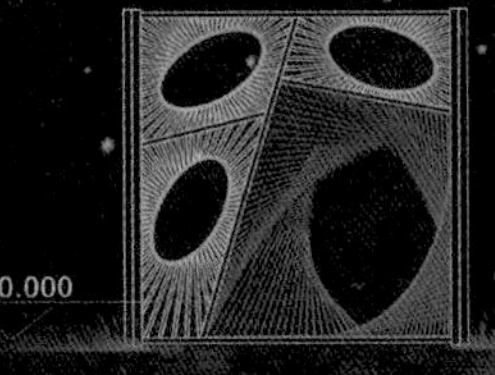

南立面图
South Elevation

北立面图
North Elevation

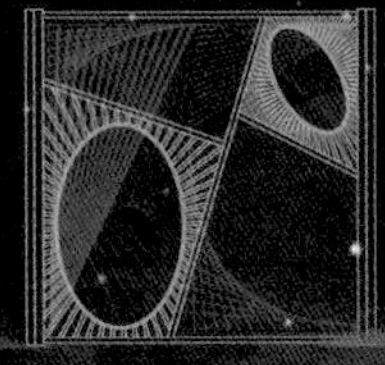

西立面图
West Elevation

0 0.5 1 1.5 m

3000
1000
600
2000
600
1000
平面图 / Plan
0 0.1 0.5 1 m

花重锦城

隐亭
INVISIBLE PAVILION

现实是物质的客观存在，
但人的精神可以挣脱现实的桎梏。

本方案以竹与草，在庸常的物质生活之外搭起一个隐秘而纯粹的角落。

在这个角落，眼睛看不到的地方，心灵可以到达，这即是生活的意义所在。希望来这里的每个人可以自我审视，自我救赎，忘记纷繁，打破枷锁，找到自我。

Reality is the objective existence of material, but human spirit can break free from the shackles of reality.

With bamboo and grass, this scheme sets up a secret and pure corner outside the ordinary material life.

Where the eyes can't see, the mind can reach, which is the meaning of life. This scheme hopes that everyone who comes here can self-examine and redeem themselves, forget the complexity, break the shackles, and find themselves.

参赛者学校：北京林业大学
指导老师：郑小东、段威
参赛者：王潇然、杨柳、薛婧、张雅迪、邓牧云、张涵林、汤大为、Wala Ali Hasan Shobil
School: Beijing Forestry University
Advisers: Zheng Xiaodong, Duan Wei
Participants: Wang Xiaoran, Yang Liu, Xue Jing, Zhang Yadi, Deng Muyun, Zhang Hanlin, Tang Dawei, Wala Ali Hasan Shobil

原型 Prototype

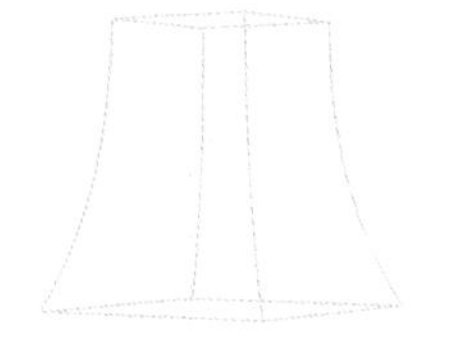

提取 Extract

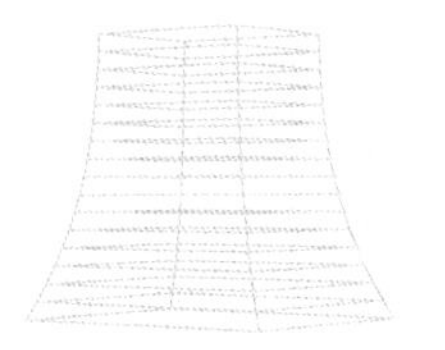

编织 Weave

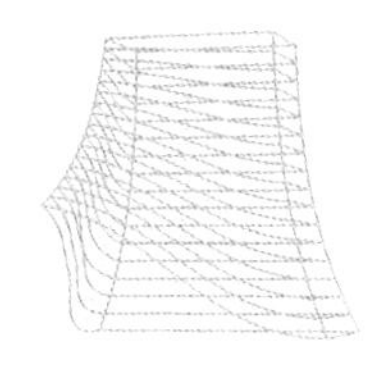

掀起 Lift

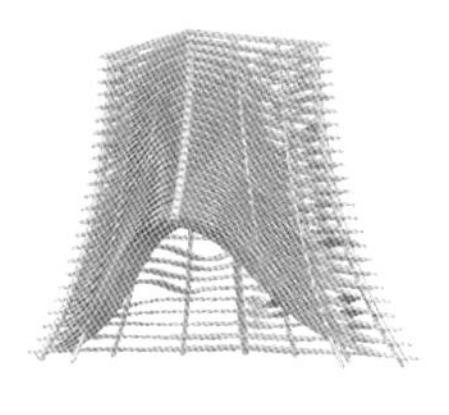

成型 Mold

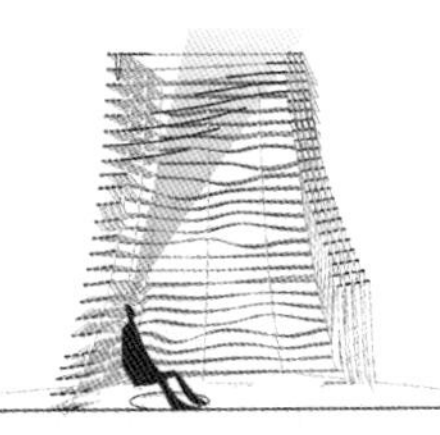

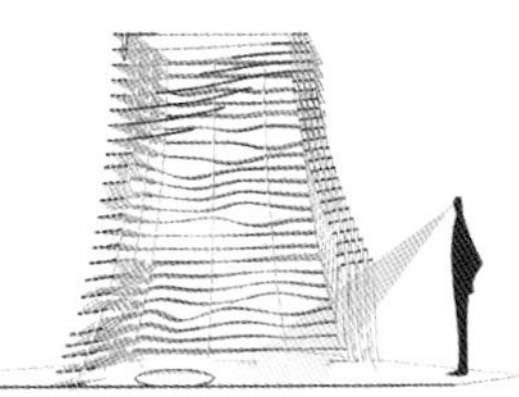

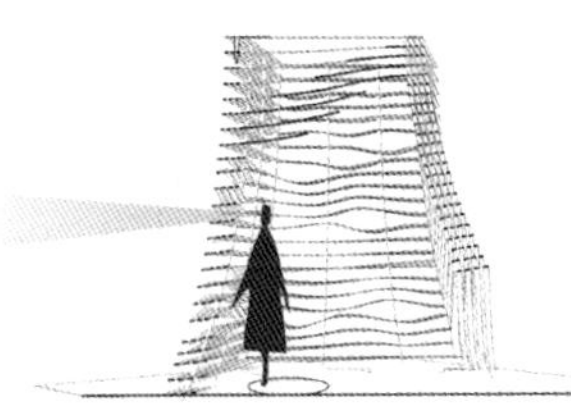

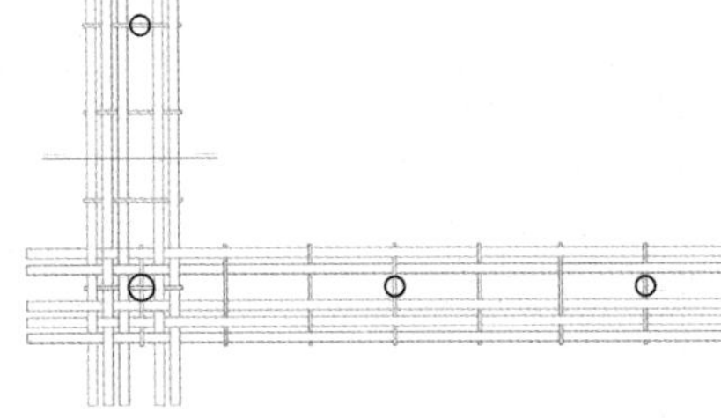

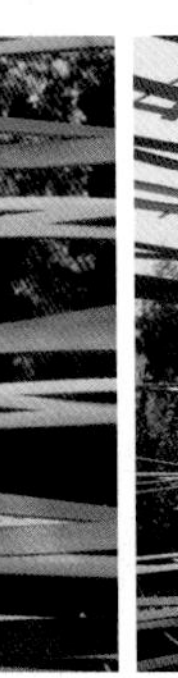

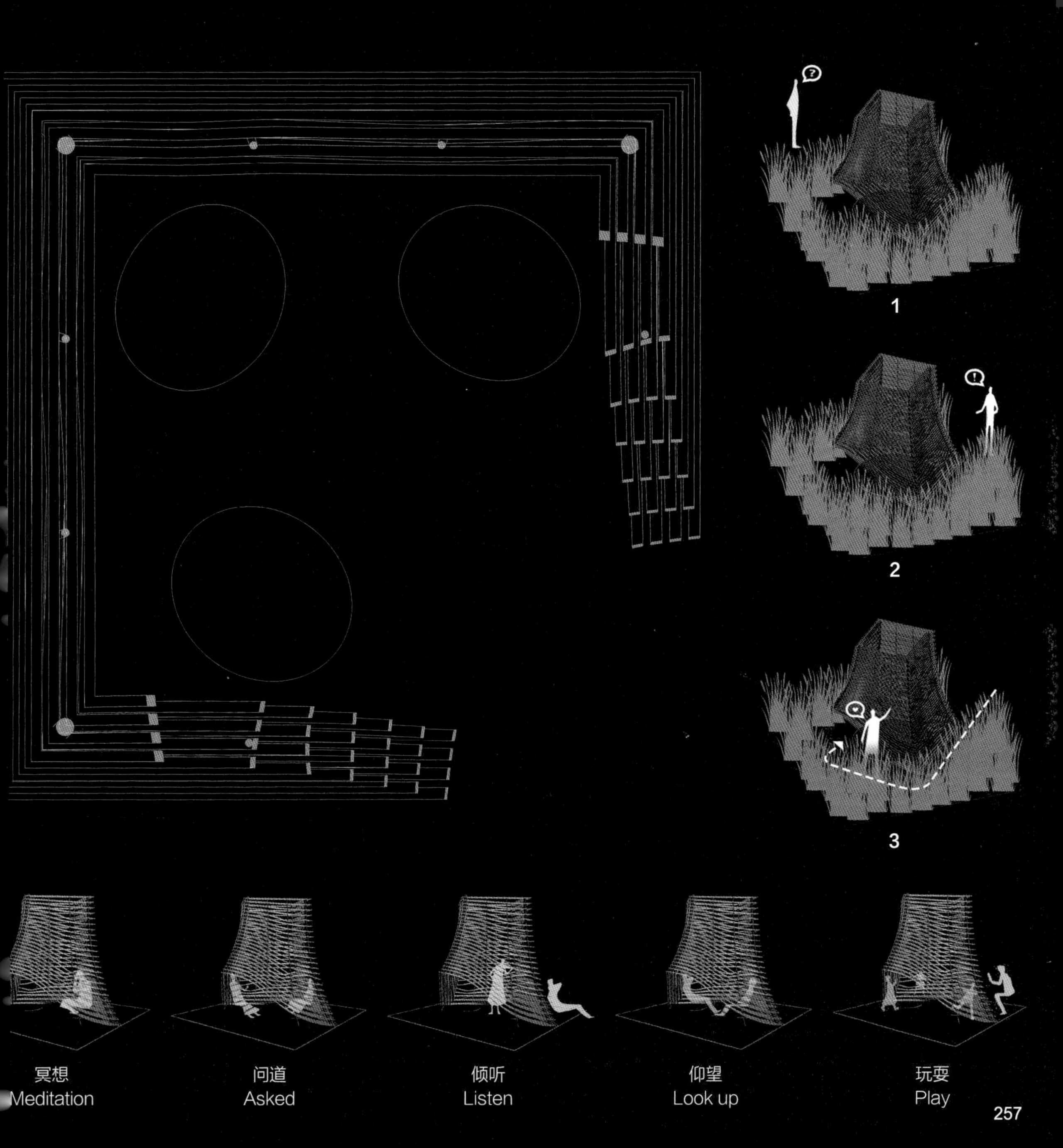
1
2
3
冥想
Meditation
问道
Asked
倾听
Listen
仰望
Look up
玩耍
Play

KFC

别有洞天
A WORLD All ITS OWN

感四时之变，
观云卷云舒。

太湖石在自然外力的洗礼下形成的孔洞无常的奇特造型，其千窍百孔中透露出神秘的美感。为了通过建筑形式表达太湖石的内部空间，我们发现了数学上的一个定义——极小曲面。极小曲面外部变化柔和，内部空间错综复杂，与太湖石瘦、透、漏、皱，曲折多变的特点相似。在经过一系列筛选和变形的过程之后，我们选取了一种相对比较常见的极小曲面——Gyroid surface 作为原型。通过参数化的设计手段，将复杂的孔洞空间浓缩为三种简易的竹构件，以自由编织的竹篾覆面，将东方的赏石文化和西方的形式科学相融合，最终营造出“别有洞天”的秘境花园。展现太湖石作为“景园要素”和“空间场所”的二重性。

Feeling the change of four seasons,
observe the clouds scudding across the sky.

Under the natural external force, Taihu stone formed a strange shape of hundreds of thousands of holes, which reveal a mysterious aesthetic feeling. In order to express the interior space of Taihu stone through architectural form, we found a mathematical definition: minimal surface. Minimal surface has soft external changes and intricate internal space, which is similar to the characteristics of thin, transparent, leaking, wrinkle and zigzag characteristics of Taihu stone.After a series of filtering and deformation processes, a relatively common minimal surface, Gyroid surface, was selected as the prototype. By parametric design, the complex cavity space is condensed into three simple bamboo components covered with freely woven bamboo strips. The oriental stone culture and the western formal science are integrated to create “A World All Its Own” and demonstrate the duality of Taihu stone as "landscape elements" and "space".

参赛者学校：北京林业大学
指导老师：段威 、郝培尧
参赛者：杨薪煜、李雪、蒋紫莹、何立飞、陈鲁、李曾莲、唐丰芸
School: Beijing Forestry University
Advisers: Duan Wei、Hao Peiyao
Participants: Yang Xinyu, Li Xue, Jing Ziying, He Lifei, Chen Lu, Li Zenglian, Tang Fengyun

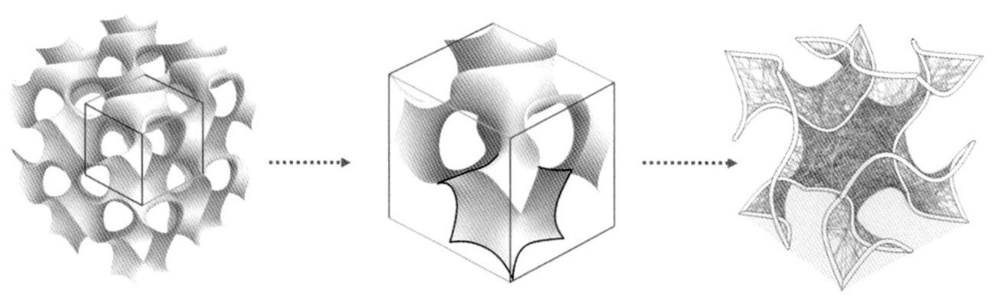

参数化设计 /Parametric design

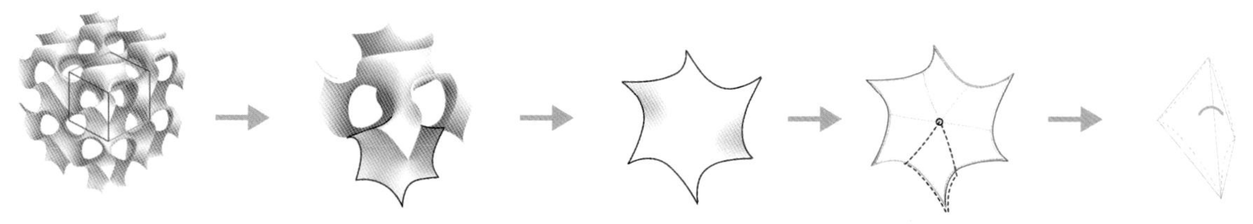

拓扑分析 / Topological analysis

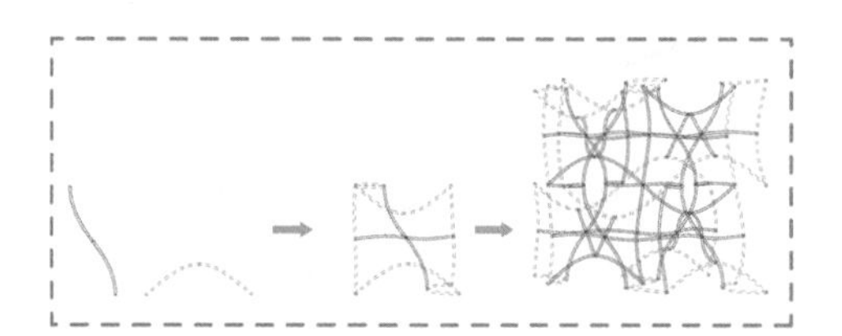

骨架分析 /Frame analysis

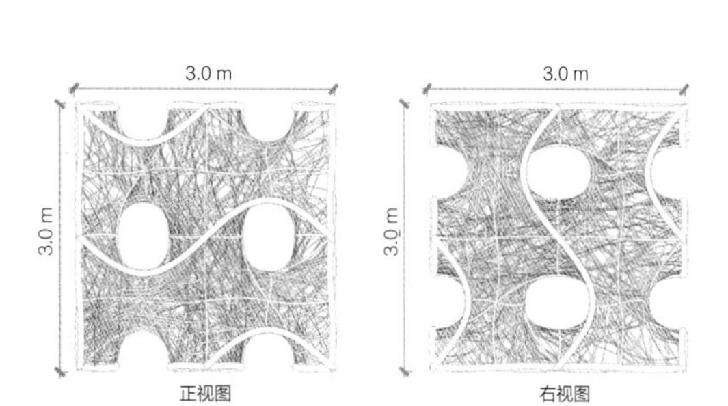

立面造型 /Elevation

瘦 / Thin

透 / Transparent

漏 / Leaking

皱 / Wrinkle

欢迎来祠洞
拍照

觅羽
FIB FEATHER IN BAMBOO

秘境如梦，觅竹如丝；丝如细羽，羽翼如飞。

秘境是浪漫的遐思，亦是斑斓的梦境。秘境应是模糊的，因此边界也是模糊的。

形态意向——“柔韧中的张力”
空间形态的张力象征着空间意向的延伸，一根根细丝仿佛束缚了空间的外延趋势，却不能束缚观者的寻觅思绪，以空间意向的无限代空间形态的有限，以无代有，隐喻性的无界空间让观者仿佛置身梦境并留有“秘境寻梦，觅竹如丝”的遐想诗意。

形态竹语——“束缚中的自由”
通过设计揭示传统竹构形态的内在材料性特征，将竹材的刚性及柔性的属性对比映射成空间形态中的张力与束缚的形态意向。正如秘境之于现实，秘境应是无限的，现实却是有限的。我们不应忘记梦境的美好，同时也无法忘却现实的约束。

The secret place is like a dream, looking for bamboo like silk, silk is like fine feathers, wings are like flying.

The secret place is not only a romantic reverie, but also a colorful dream. The secret place should be fuzzy, so the boundary is also fuzzy.

Form Intention — "tension in flexibility"
The tension of space form symbolizes the extension of space intention. The filaments seem to constrain the extension trend of space, but they can't constrain the viewer's searching thoughts. The infinite generation of space intention, the limited generation of space form, and the metaphor of unbounded space make the viewer feel as if he were in a dream and left with the reverie of "seeking dreams in a secret place, looking for bamboo like silk".

Material Intention — "freedom in bondage"
Through the design, the inherent material characteristics of traditional bamboo structure are revealed, and the comparison between the rigid and flexible properties of bamboo is mapped into the form intention of tension and bondage in the spatial form. Just as the secret realm is to the reality, the secret realm should be infinite, but the reality is limited. We should not forget the beauty of dreams, but also the constraints of reality.

参赛者学校：上海交通大学
指导老师：陈霆、张洋、于冰沁
参赛者：崔自佳、陈宇钢、张恩嘉、曹一丹、吴佳远、赖佳妮、汤颖慧、杨清枫
School: Shanghai Jiaotong University
Advisers: Chen Ting, Zhang Yang, Yu bingqin
Participants: Cui Zijia, Chen Yugang, Zhang Enjia, Cao Yidan, Wu Jiayuan, Lai Jiani, Tang Yinghui, Yang Qingfeng

寻幽 Seeking

入境 Coming

羽升 Flying

见羽 Seeing

人群活动 /People activities

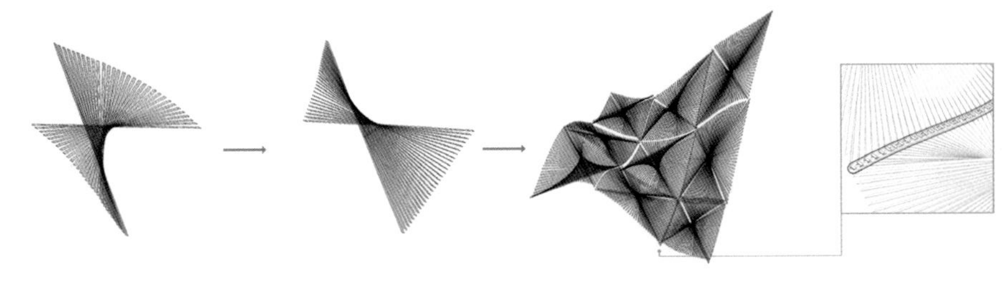

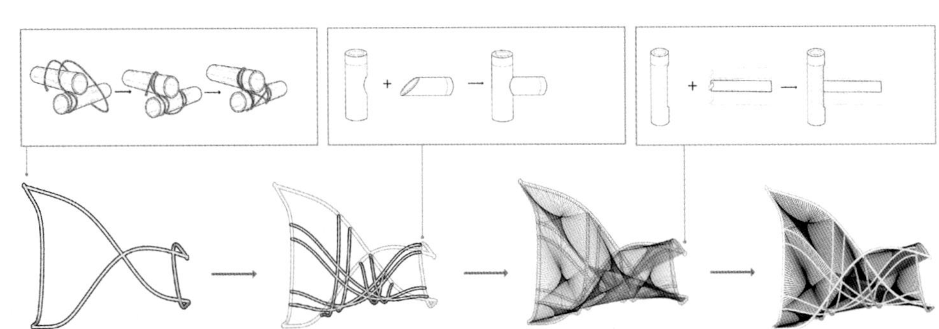

结构 /Structure

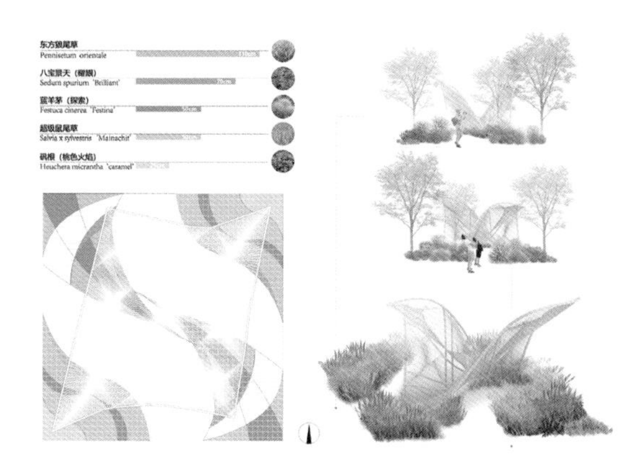

种植设计 /Planting design

蜀韵幽景

SHU YUN YOU JING

蜀韵——
是文人墨客与雅士对精神空间以及园林空间的最高追求，意境洒脱悠然。
幽景——
一幅清幽自然的竹林山水画卷，山川河谷之间凉风习习，竹叶飘飘，万物繁荣，生机勃勃。

川派盆景的姿态、川西林盘的意境和西蜀园林的幽景都浓缩在整个竹构架中，使观者能够感受蜀地自然与人文环境的浸润与熏陶。
青城的幽、峨眉的秀、西岭的峻、剑门的险，统统浓缩在顶端蜀山的蜿蜒不绝之中。

参植美竹异卉，荟翳参差，而春芳夏阴，波光月晖，以时献状无不可爱，故为成都园亭胜迹之最。

Shu Yun — The charm of Shu
It is the highest pursuit of literati and scholars for spiritual space and garden space, with free and easy artistic conception.
You Jing — The serene scene
A quiet and natural landscape painting of bamboo forest, with cool breeze and bamboo leaves fluttering between mountains, rivers and valleys. All things are thriving and full of vitality.

The posture of Sichuan school bonsai, the artistic conception of western Sichuan forest plate and the secluded scenery of western Shu garden are concentrated in the whole bamboo frame, so that visitors can feel the natural and cultural environment of Shu.
The seclusion of Qingcheng, the splendour of Emei, the steep of Xiling, and the danger of Jianmen are all concentrated in the winding of Shushan Mountain at the top.

Elegant bamboos and flowers form a shelter, while the flowers in spring, the shade in summer and the light in the moon are all lovely, so they can form the most famous garden pavilion in Chengdu.

参赛者学校：四川农业大学
指导老师：陈其兵、郭丽
参赛者：谢守红、杨史前、姜宇謦、畅雨婷、唐可柚、龙傲宇、张姮
School: Sichuan Agricultural University
Advisers: Chen Qibing、Guo Li
Participants: Xie Shouhong, Yang Shiqian, Jiang Yuxin, Chang Yuting, Tang Keyou, Long Aoyu, Zhang Heng

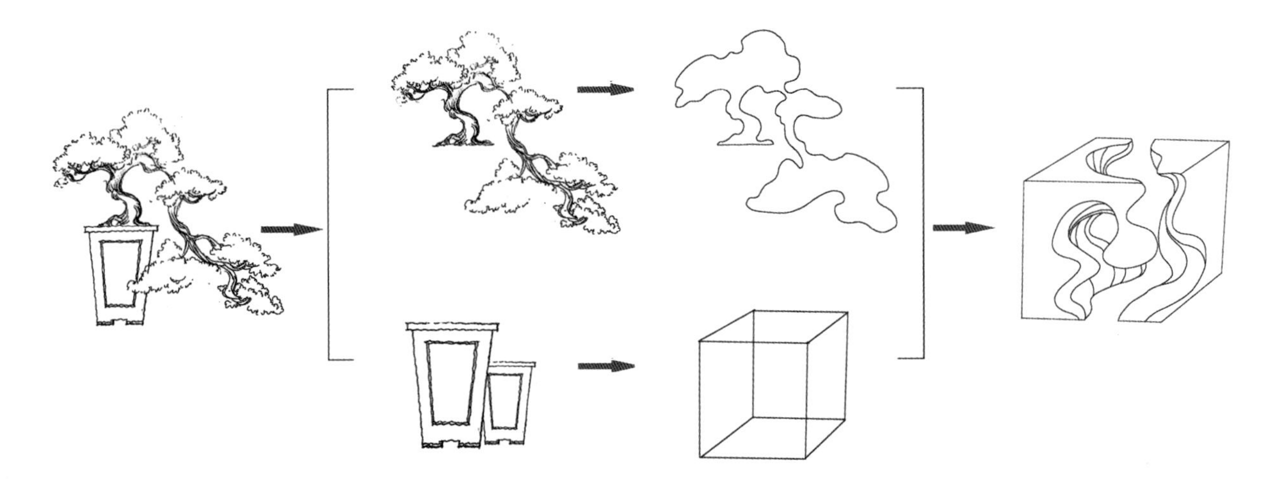

生成 /Create

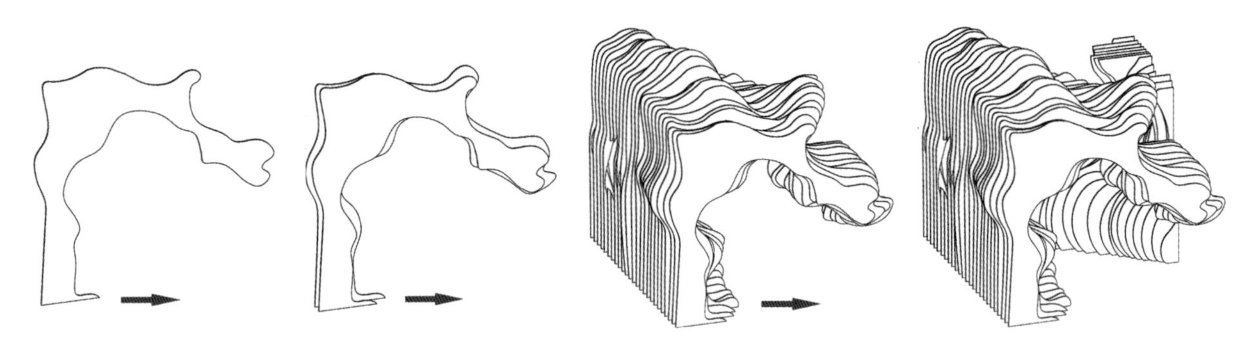

结构 /Structure

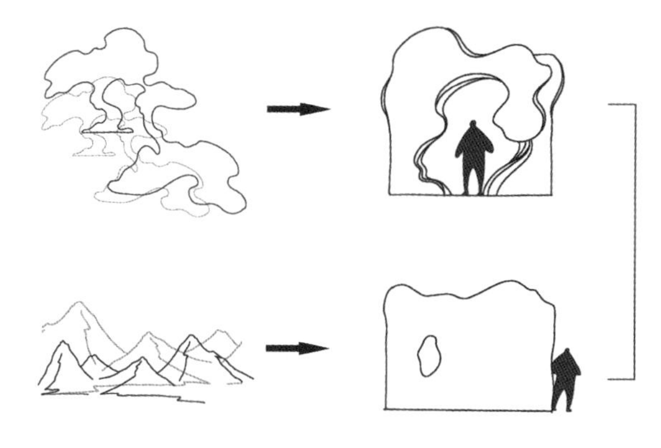

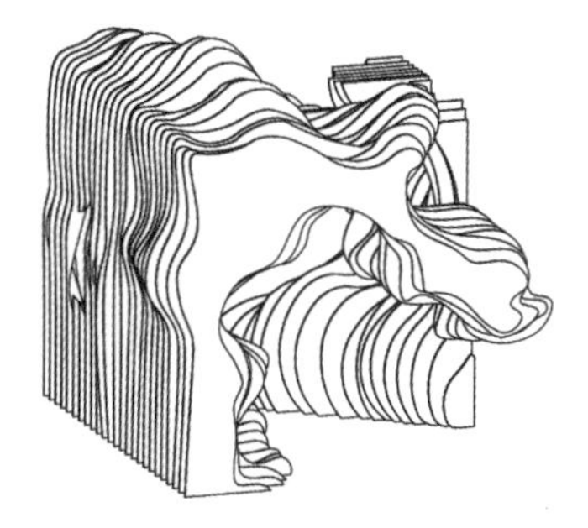

概念 /Concept

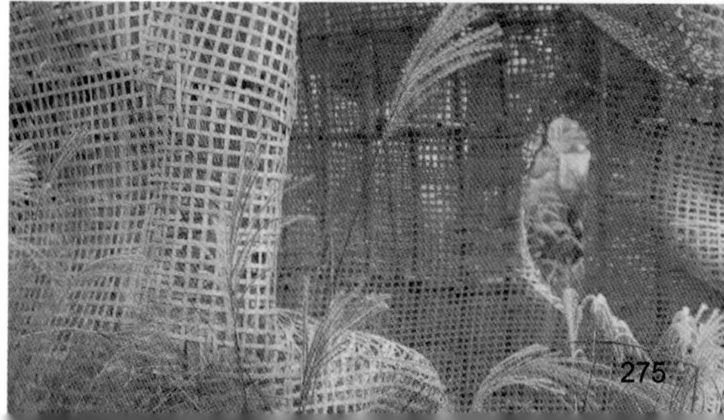

尘垣
MEDITATION

方案灵感来源于风蚀走廊，透过节理复杂的层岩、错落有致的地形，如同勘探高深莫测的秘境，结合了风蚀走廊和竹条的共同点，在色彩、编制方面形成天然、朴素的艺术特色，形成“尘垣”的设计方案。

方案通过两个角度、三面围墙的围合，表现出风蚀遗留的岁月蹉跎，层叠销蚀的躯壳象征世间纷扰，内在木雕象征着人的静思。在浮躁的现代生活中，人们在光影变幻莫测的竹构物中静思冥想，远离尘嚣，从而追求内心的平静。在经历学习和工作的劳累后，享受原始的本真本色，无匮乏无享受，无所求无所惧。

本设计旨在展示静思与现代的联系，为现代人与安宁的善美境界之间提供联结的可能。心情宁静地进行冥想与沉思，以静，寻求人体最本质的状态；以观，在本质状态下，找到最核心的源头。内心的淡定从容，定而后能静，静而后能思，思索自身的本源，得到内心的自由。设计把风沙与沉思化为静物，体现风沙在千百年的时光中风化成瑰丽沧桑的秘境，令游人沉静，给人以心的滋养。

Our idea comes from wind-erosion landform. Walking through the rock formation with complex cleft, you will feel like exploring mystical neverland. We were going for a natural and simplistic aesthetic which draws attention in part to the colors and weaving techniques.

Our design not only showcases the similarities between bamboos and rocks, but also some spiritual symbols. Two angels and three walls show the years gone by. The wind left nothing, but the fabulous shape of the rock. The woodwork is like an old man sitting on the stone in deep thought. One must keep away from the ash and the dust, especially in such confusing modern times. People will meditate in the bamboo-area after being so tired of working, just to have an area silence and enjoyment. The nature reflects itself with no desire.

Our design shows the relationship between modern life and rumination, connecting the people with a wonderful state of silence. Slowly go down, try to find the shape of humanity's raw beauty by mediating. Observe not only the grief but also their own inner-being as well. If you wish for mental freedom, I suggest that you calmly explore the corners of your mind and most importantly find it within yourself. Wind, sand, and rumination will be the focus of the design. People will meditate the fairy like never land only to awaken their natural soul.

参赛者学校：西北农林科技大学
指导老师：李志国
参赛者：徐艺峰、张乐其、王逸涵、温静怡、范世玉、杨可欣、张茜、丁枭
School: Northwest A&F University
Advisers: Li Zhiguo
Participants: Xu Yifeng, Zhang Leqi, Wang Yihan, Wen Jingyi, Fan Shiyu, Yang Kexin, Zhang Qian, Ding Xiao

编织纹理 /Woven texture

编织纹理 /Woven texture

墙面纹理 /Wall texture

形态推演 /Form generation

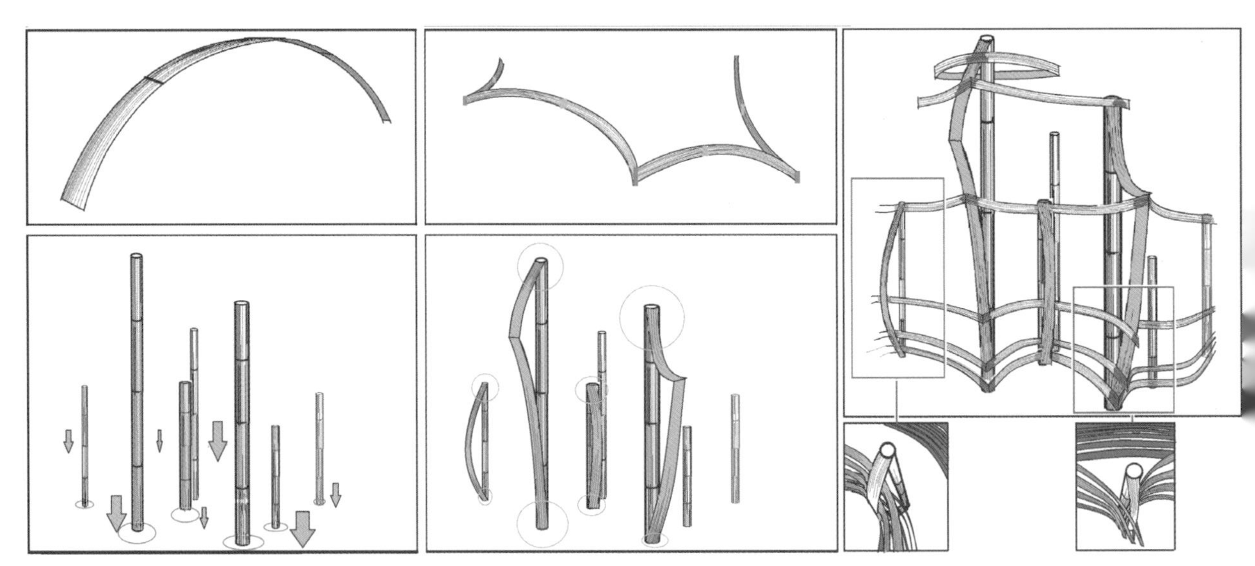

建造方式 /Construction process

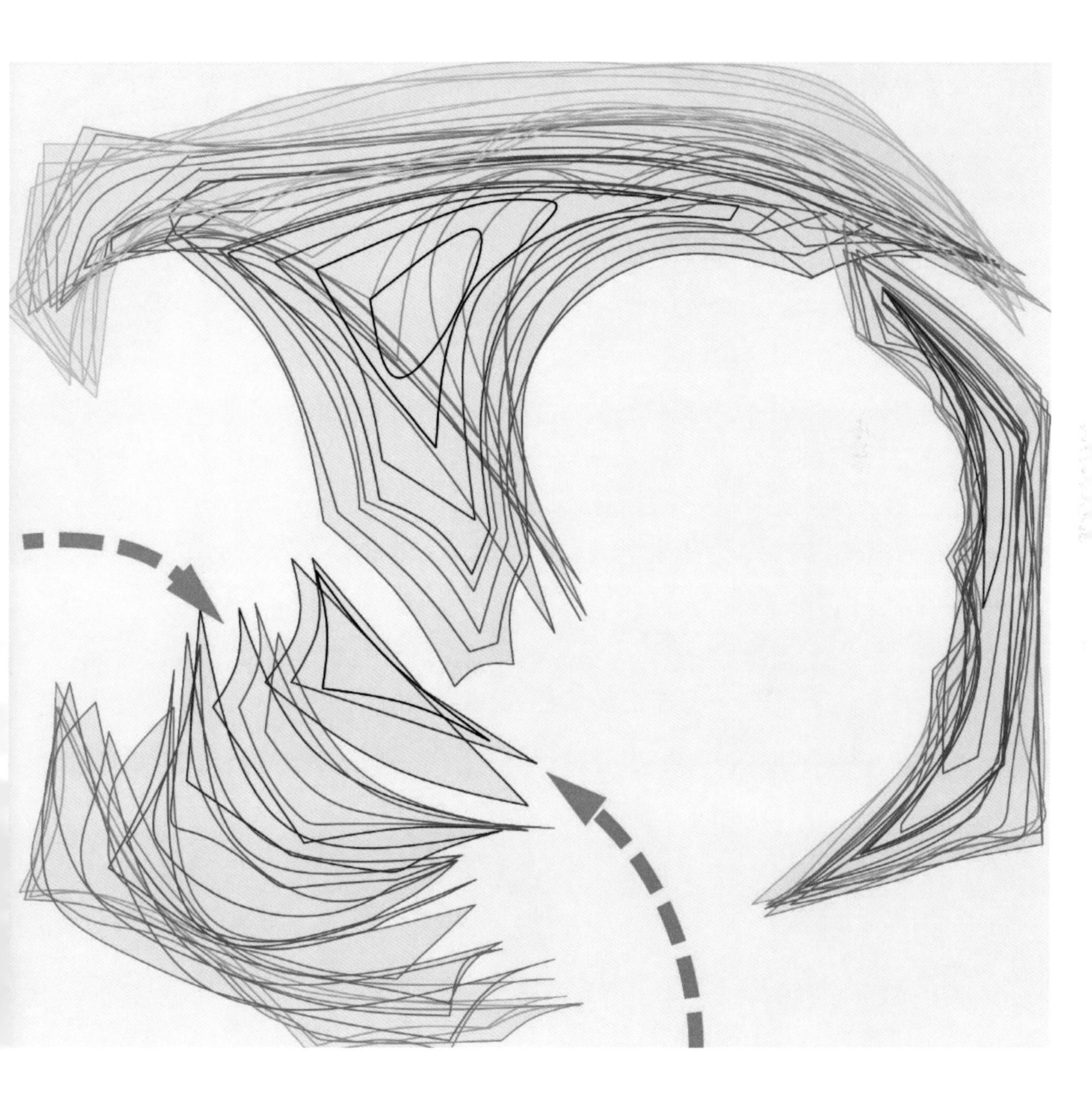

一方·竹里
THE CORNER OF BAMBOO · THE CUBE OF HORIZON

从唐代诗人王维的《竹里馆》汲取设计内涵，从基础的方形空间入手，运用现代的景观语言，表达古代意境之美。用可参与的动态构件，营造丰富的内部空间体验；用简洁易行的构造方式，使其便于落地建造。凸显了可变、可赏、可游、可坐，一同创造了多元的竹里空间，将“竹深无人知，唯与竹声伴”的氛围融入秘境花园创作。

This scheme draws the design connotation from the "Zhuliguan" (the Villa in Bamboos) by Wang Wei, a poet of the Tang Dynasty. The project starts from the basic square space and uses modern landscape language to express the beauty of ancient artistic conception. Dynamic components that can participate are used to create a rich internal space experience. With a simple and easy construction method, the pavilion is easy to build on the ground. It highlights the changeable, admirable, walkable, and stayable space, and creates a diverse bamboo space together, integrating the atmosphere of "bamboo is deep in no one knows, but with the sound of bamboo" into the secret garden creation.

参赛者学校：华中科技大学
指导老师：苏畅、戴菲
参赛者：孙培源、丁璐、裴子懿、王佳峰、邱悦、蔡卓霖、李姝颖、杨超、赵广旭
School: Huazhong University of Science and Technology
Advisers: Su Chang, Dai Fei
Participants: Sun Peiyuan, Ding Lu, Pei Ziyi, Wang Jiafeng, Qiu Yue, Cai Zhuolin, Li Shuying, Yang Chao, Zhao Guangxu

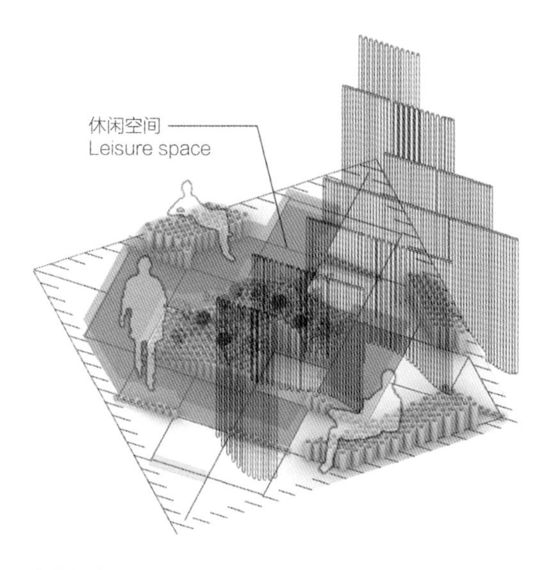

余荫栖息 /Halt

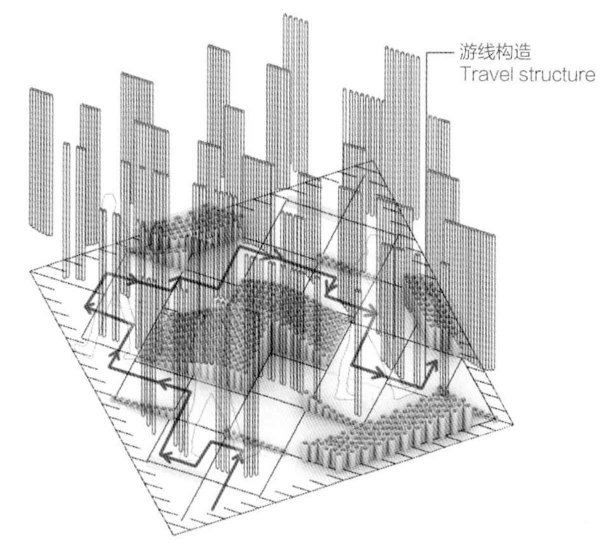

千尺万竿 /Pathway

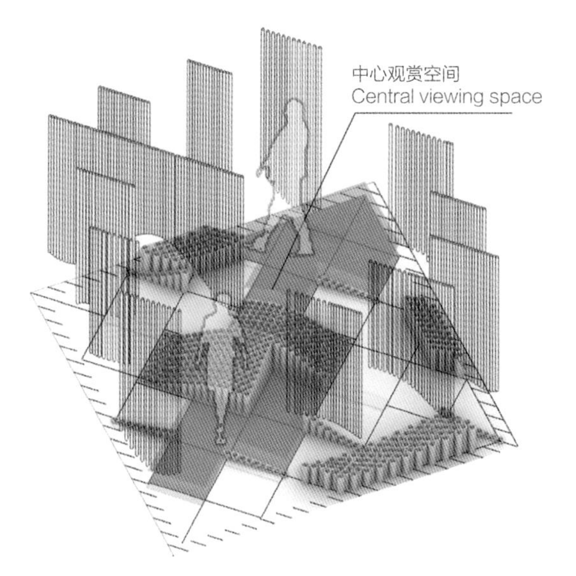

明画千寻 /Watch

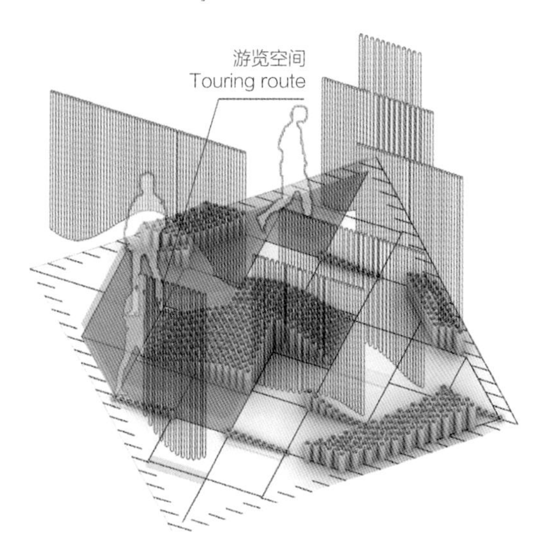

幽林抽荫 /Stroll

“zhu” 梦蜀境
"ZHU" DREAMS IN SICHUAN FAIRYLAND

巴蜀佳酿，历史悠久，源远流长。川酒承载着巴蜀文化，也给予了四川人逍遥和自在的世俗精神。它是你悠闲时的“浅把涓涓酒，深凭送此生”，也是你忧愁时的“蜀酒禁得愁，无钱何处赊”。它能让你物我两忘，物我同一，把你带至心中的秘境。

“zhu”梦蜀境中的“zhu”同“逐”“竹”“筑”，我们追逐梦想，用竹子筑造巴蜀秘境。单个竹构由多条曲率相同的竹条相互交错形成外框，表皮呼应交错的竹条构成疏密变化的菱形编织，前后由形似掀开的竹帘构成出入口，内部由中心的竹筒和环绕的竹圈起结构支撑作用。三个竹构形似酒瓶错落摆放，飘带串联彼此，形成三人对饮，酒香环绕的情境。植物配置呼应竹构曲线，呈高低有致、自然环绕分布，增添秘境氛围。竹构之间萦回婉转，空间多变，你可以独自“举杯邀明月”，仰望天空星河灿烂；也可以与知己“欢言得所憩，美酒聊共挥”。

The exquisite Sichuan vintage alcohol enjoys of long history and carries on Ba and Shu culture, as well as cultivates Sichuan people's leisurely and unrestrained spirits. It could be either accompany with rest of your leisure life, or consolation when you are in bad mood. It can make you forget all and bring you to the secret place at the bottom of your heart.

"Zhu" Dreams in Sichuan Fairyland, where "zhu" is equivalent to chasing, bamboo and construction in Chinese homophone, illustrating that we chase our dreams to create the Sichuan fairyland with bamboo. The outer frame is composed of several bamboo strips with the same curvature. The skin, corresponding to a diamond shaped braid is weaved with diamond strips. The entrance and exit are made up of bamboo curtains. The inner part is supported by the central bamboo tube and the surrounding bamboo ring. The three bamboo structures which are like alcohol bottles and connected with the ribbons are in series to form a situation where three people drink with each other, and the fragrance of alcohol is surrounded. The plant configuration echoes the bamboo structure curve and demonstrate tall and low rhythm and secret atmosphere. Here you can either "raise your glass to invite the bright moon and stars" or "have a rest with your bosom friend and have a good drink".

参赛者学校：重庆交通大学
指导老师：罗融融、温泉
参赛者：宋雨芮、杨涛、肖佳妍、夏宇晨、张沂珊、牟鹏、邱雅雄、刘淘孟
School: Chongqing Jiaotong University
Advisers: Luo Rongrong, Wen Quan
Participants: Song Yurui, Yang Tao, Xiao Jiayan, Xia Yuchen, Zhang Yishan, Mou Peng, Qiu Yaxiong, Liu Taomeng

概念生成 /Concept generation

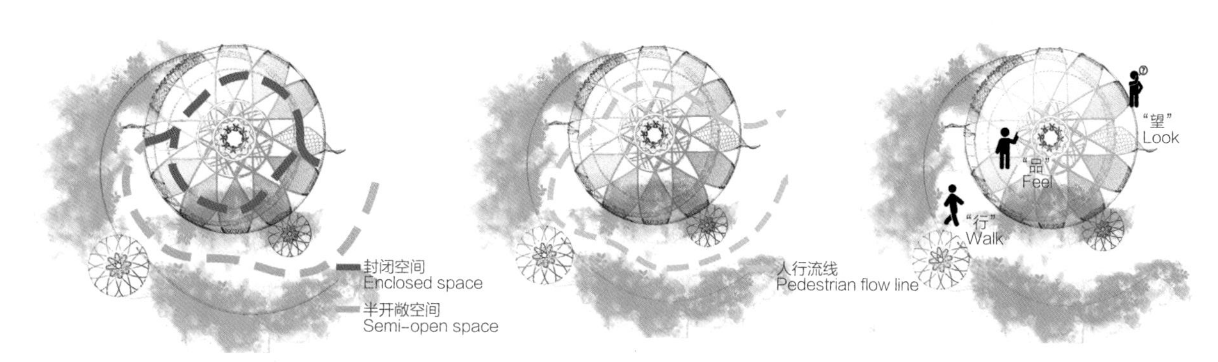

空间与功能分析 /Space and function analysis

立面图 /Elevation

未知
THE UNKNOWN

世间万物均由原子构成。方案中，我们选择用抽象的球体作为组成世界的单元，并通过球面丰富的形态变化赋予球体更多元的含义。

当你置身于构筑之中，你如同穿梭在一个个神秘的未知世界里，你可以选择俯身把玩，可以选择矗立观望，甚至可以改变球体的位置，创造新的世界……

它或是黑夜中明亮的灯群，或是漫天繁星，抑或是翻腾的海洋，每个球体、每个空间都因你我想象的不同而生出丰富的形象。你我穿梭其中，你永远不知道我为你留下的是何样世界。

我们笃信，秘境的真谛在于永恒的未知，以及它所带来的期待与想象。

Everything in the world is made of atoms. In our design, we regard the abstract ball as the unit of the world and endow it with full meanings through its rich morphological changes.

When you are in the structure, you are like shuttling through a mysterious unknown world. You can choose to bend over to play, to stand and see, and even change the position of the polysemy balls to create a new world...

Maybe it is a cluster of bright lights in the dark night, or the stars in the sky, or the churning sea. Every ball, every space, is enriched by the difference between our imagination. Wandering in it, you never know what kind of world we leave for you.

We firmly believe that the essence of mystery is the eternal unknowns, and the expectations and imagination it brings.

参赛者学校：华南理工大学
指导老师：谢纯
参赛者：黄海琪、徐琦、何静、马晓旭、杨扬、潘麒羽、梁杰麟、刘宇嘉、木云童、何佩琪
School: South China University of Technology
Advisers: Xie Chun
Participants: Huang Haiqi, Xu Qi, He Jing, Ma Xiaoxu, Yang Yang, Pan Qiyu, Liang Jielin, Liu Yujia, Mu Yuntong, He Peiqi

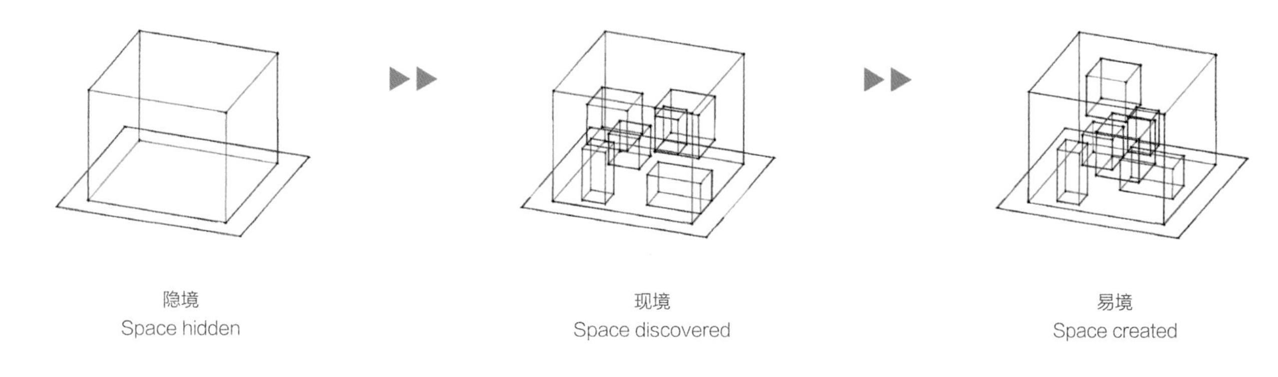
隐境
Space hidden
现境
Space discovered
易境
Space created

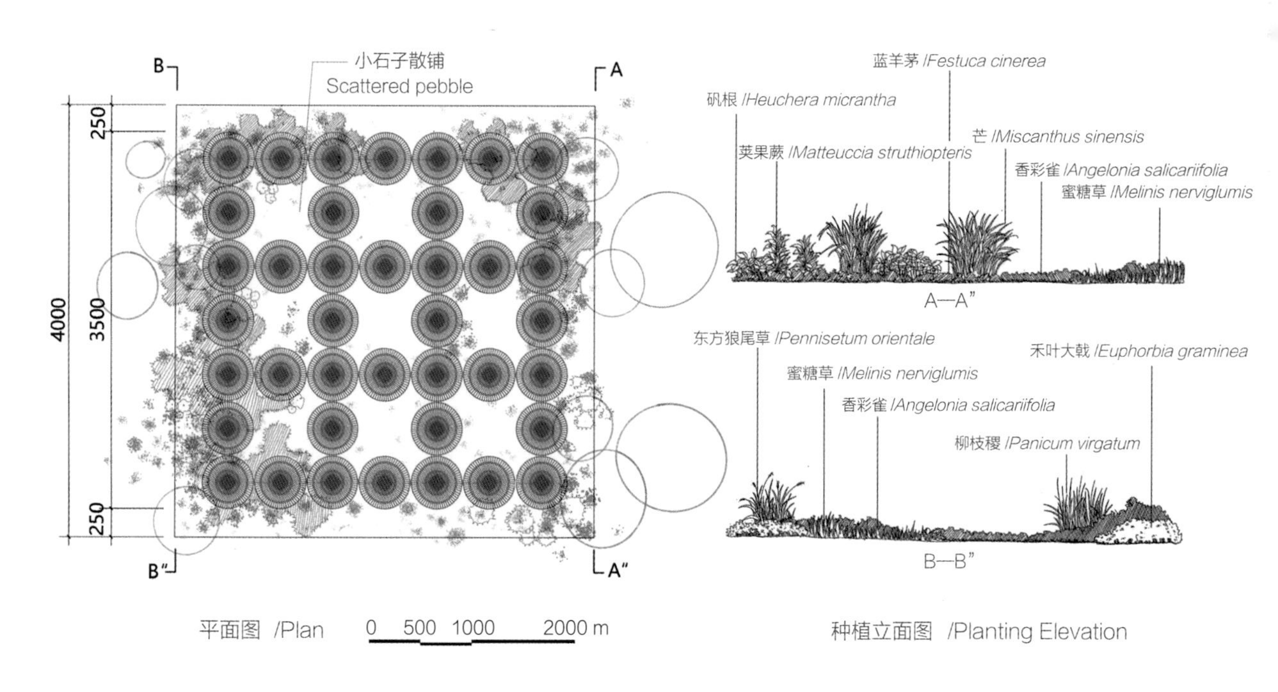
B
A
小石子散铺
Scattered pebble
250
4000
3500
250
B"
A"
平面图 /Plan
0 500 1000 2000 m
蓝羊茅 /Festuca cinerea
矾根 /Heuchera micrantha
芒 /Miscanthus sinensis
荚果蕨 /Matteuccia struthiopteris
香彩雀 /Angelonia salicariifolia
蜜糖草 /Melinis nerviglumis
A—A"
东方狼尾草 /Pennisetum orientale
禾叶大戟 /Euphorbia graminea
蜜糖草 /Melinis nerviglumis
香彩雀 /Angelonia salicariifolia
柳枝稷 /Panicum virgatum
B—B"
种植立面图 /Planting Elevation

轮回之境
THE REALM OF SAMSARA

轮回

以佛教论，众生因对世间无常的真相无所了知，或因对生命的实相不明了，而导致不断生死的烦恼未断，便在六道中如车轮一样地旋转，即“六道轮回”。拓扑变换中不可变换点交错重合，环体现封闭循环性，使之轮回。

轮回之境

本方案将两个方向垂直的环体交错形成空间，形成丰富的视角，仿佛置身于时空隧道。人走进时空隧道，单侧环状不可定向，如同拓扑变换中不可遇见的两点，在交错时空中重合，任意变形下保持不变的结构，意味着永恒未必不可接近。

在真实与想象的秘境世界里，寻找认同，发现自我。正如两面即一面的莫比乌斯环，人心的矛盾也在此对立统一，没有世人说不清的正反，只有内心走不完的轮回。

Reincarnation

According to Buddhism, sentient beings constantly worry about life and death, because they are ignorant of the impermanent truth of the world, or they are not aware of the reality of life. So, they spin like wheels in the six paths, which is called "the six paths of samsara". In topological transformation, the incommutable points are staggered and overlapped, and the ring reflects the closed circularity, making it reincarnation.

The realm of reincarnation

This scheme makes two rings that are perpendicular to each other to form a space, forming rich visual angle, as if place oneself in the tunnel of time and space. People walked into the tunnel of time and space, single ring is directional. Like two invisible points in topological transformation, the overlapping of time and space and the structure that remains unchanged under arbitrary deformation means that eternity is not necessarily inaccessible.

In the secret world of the real and the imaginary, we seek for identity and find ourselves. Just like the Mobius ring of two sides being one side, the contradiction of the human heart is also in this unity of opposites. There are no pros and cons that people can't tell, but only endless reincarnation in the heart.

参赛者学校：西华大学

指导老师：曹伦、钟锦玉

参赛者：许立顺 李竟语 徐浩可 吴悠 王茜 乔巧 周玉双 王丽萍

School: Xihua University

Advisers: Cao Lun, Zhong Jinyu

Participants: Xu Lishun, Li Jingyu, Xu Haoke, Wu You, Wang Qian, Qiao Qiao, Zhou Yushuang, Wang Liping

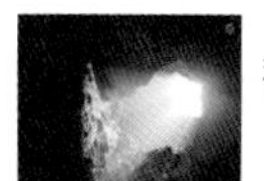

秘境 Shakotan coast

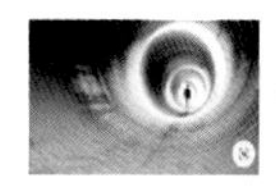

时空 Spacedme

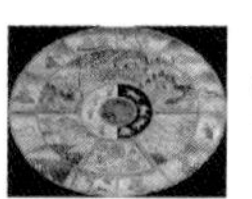

轮回 Reincarnation

概念生成 /Concept generation

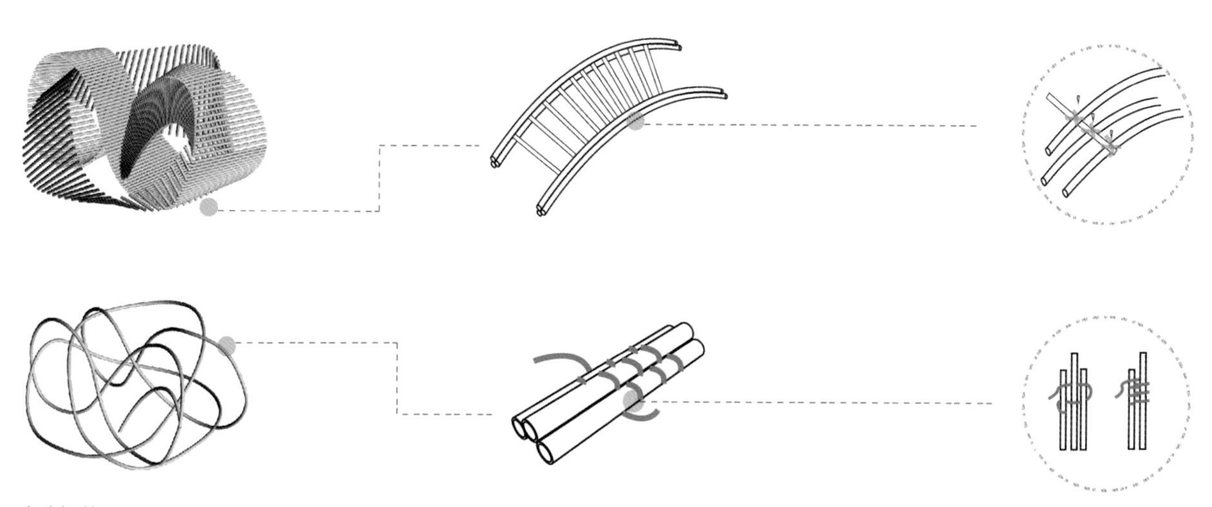

建造细节 /Structural details

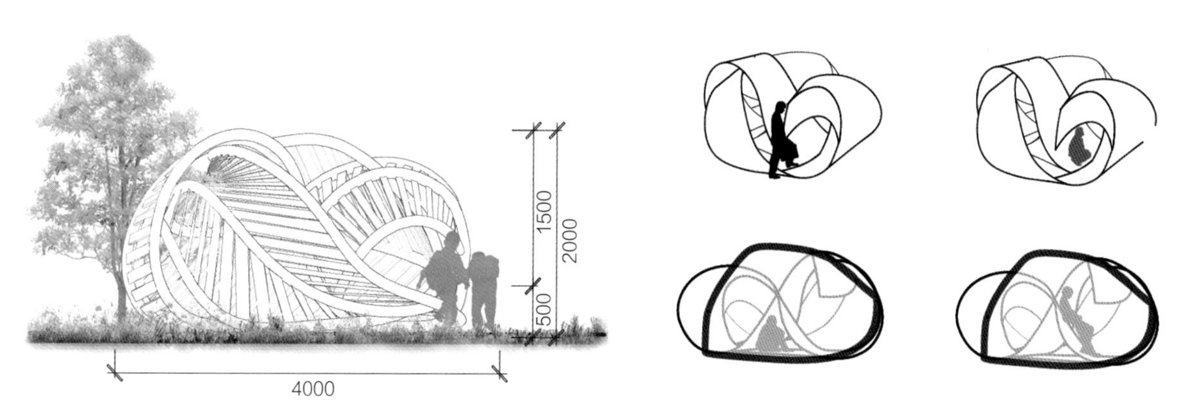

立面图 /Elevation

行为活动 /Activities

梦·蝶

DREAMING BUTTERFLY

“昔者庄周梦为蝴蝶，栩栩然蝴蝶也，自喻适志与！不知周也。俄然觉，则蘧蘧然周也。不知周之梦为蝴蝶与？蝴蝶之梦为周与？”

庄子认为，不同事物之间是可以相互转化的。其表达的不仅是物与我之间的本真关系，同时也是借由我的主体表达物与物之间的物化转变。

竹构设计整体采用双层结构，内部形似蝴蝶展翅，外部形成扭曲不稳定的空间形态。通过内外的两重形态，在变与不变间转换，激发游览者对“秘境”花园的无限想象。

那是谁在变化？在互相转化中，究竟何为蝴蝶，何为庄周？

俄而觉醒，蝶与境兀自构成一幅缱绻绚丽的梦境。

"Zhuangzhou dreamed of himself becoming a butterfly, floating and chill on air. He totally forgot he was Zhuang Zhou at that moment. After a while, he woke up, curiously wondering about if he was Zhuang Zhou himself. Then he immersed himself into serious thinking, but he still did not know whether Zhuang Zhou dream into a butterfly, or the butterfly dream into Zhuang Zhou?"

According to Zhuang Zi, different things can be transformed into each other. We can learn from it that the core of philosophy of Laozhuang is not only the true relationship between things and me, but also the materialistic transformation between things through the subject of "me".

With the double-layer structure, the outer epidermis looks like a disordered change, and expresses the sense of stability in an unstable state, while the internal stable form with a straight-grain surface creates different forms in different angles, forming a sense of instability in the space. Through the dual forms of internal and external, the transformation between change and immutability is our imagination of the secret land.

As for the overall shape, the internal shape resembles the wings of a butterfly, and the outer shape is distorted and unstable. Through the two forms of inside and outside, changing between change and unchanging, this work stimulates visitors' infinite imagination of the "Secret Land" garden. So who is changing? In the process of mutual transformation, what is the butterfly, and what is zhuangzhou?

参赛者学校：成都理工大学、南京师范大学中北学院
指导老师：徐巧 莫妮娜
参赛者：杨弋渟、雷语、余柯瑶、王嵬生、潘攀、金凤
School: Chengdu University of Technology, Nanjing Normal University Zhongbei College
Advisers: Xu Qiao, Mo Nina
Participants: Yang Yiting, Lei Yu, Yu Keyao, Wang Weisheng, Pan Pan, Jin Feng

梦 · 蝶

曲竹 · 取径 · 趣山

BENT BAMBOO · MYSTERIOUS ROAD · ROLLING HILLS

本方案概念源于《桃花源记》中“初极狭，才通人，复行数……豁然开朗”的场景，提取出“探索 + 流动”作为契合主题秘境花园的空间。再从以意化境的山水画中，抽取起伏的山峦作为竹构的形态。结合人的行为，最终生成竹构形态。在结构与材料的应用上，使用竹片加圆竹为骨架，竹片加竹篾为表皮。种植设计方面，整体取花青的蓝绿色调，远观显悠然，内部则点缀活泼的绿色。花期集中在 6~9 月，四季有景，各有风味。

The concept of the scheme comes from the *Land of Peach Blossoms*, "The cave is very narrow at first, and only one person can pass. Going further dozens of steps, the scene suddenly becomes spacious." This scheme extracts "exploration + flow" as a space that fits the theme of the secret garden. Then, the undulating hills are extracted from the landscape paintings to represent the landscape. In combination with human behavior, bamboo structure forms are finally formed. In the application of structure and materials, bamboo slices and round bamboo are used as the skeleton, and bamboo slices and bamboo strips as the skin. In terms of planting design, the structure overally uses flowers of blue-green tone, which makes the distant view appears leisurely. And the interior is botted with lively green. Flowering in June to September, the four seasons have scenery, each has its own charm.

参赛者学校：西安建筑科技大学
指导老师：武毅
参赛者：赵虎宸、朱莉梅、徐保平、罗伍春紫、王育辉、谢欣阳、马樱宸、李康华
School: Xi'an University of Architecture and Technology
Advisers: Wu Yi
Participants: Zhao Huchen, Zhu Limei, Xu Baoping, Luowu Chunzi, Wang Yuhui, Xie Xinyang, Ma Yingchen, Li Kanghua

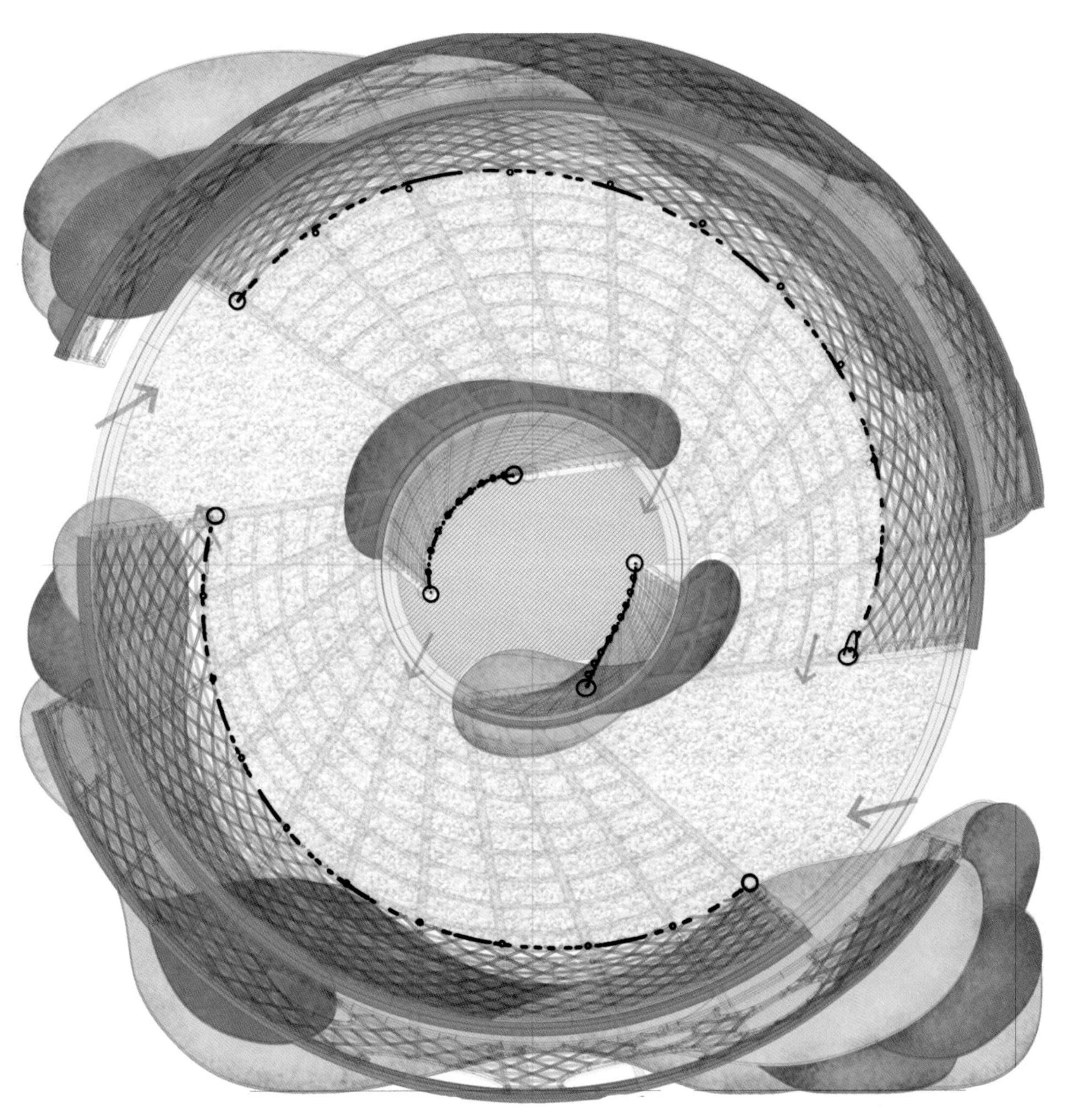

玉簪
Hosta plantaginea
天门冬科 玉簪属

觅水影
FOLLOW THE SHADOW, WANDER IN THE RIPPLE

"蒲草青青苇色新，阳光逐水水生金。"阳光透过摇曳的芦苇秆，在水面上洒下了金色的微波。我撑起小小的渔船穿梭在丛丛芦苇间，追着光，觅着水影，缓入神秘浪漫的苇丛深处。

秘境是寻常而亲近的，是在我们平常的生活中，等待着被感知和发现的。

设计以"觅水影"为题，通过模拟水波的编织方式，再现了苇丛中水与光交融的场景，借此期许人们能够发现寻常生活中的美与神秘。蜿蜒的路径划分出三处小花园，丰富的空间变化叙述着探寻秘境的四个阶段——窥、寻、思、品。植物主要选用竖线条的观赏草，营造芦苇荡中自然野趣的氛围。色彩上借用了印象画派的"光色技法"，采用原色并列、重叠和补色手法配置植物的色彩，淡雅的花色在阳光下形成了细微的颜色变化，形成极具氛围感的光影印象，点"水影"之题。

The sunlight sprinkled golden shimmer on the water through the swaying reed stalks. I propped up the small fishing boat to shuttle between the bushes and reeds, chasing the light, looking for the water shadow, and slowly descending in the depths of the mysterious and romantic reeds.

The theme of the design is "Follow the shadow, Wander in the Ripple", which reproduces the scene where the water and light blend in the reeds by simulating the weaving of water waves. The space composed of winding path and a small garden foreshadows the four stages of exploring the secret realm, namely peeping, searching, thinking, and tasting.

The plants are mainly ornamental grasses with vertical lines to create a natural and wild atmosphere in the reed marshes. In terms of color, it borrows the "light-color technique" of impressionism painting. The colors of plants are arranged by juxtaposing, overlapping, and complementary colors of primary colors. The elegant flower colors change slightly under the sun, forming a very atmospheric light and shadow impression, thus emphasizing the topic of "water shadow".

参赛者学校：北京林业大学
指导老师：王向荣、赵晶
参赛者：张沚晴 刘琦 马琳 吴沿羲 吴雨轩 蒋鑫 李见哲 霍达
School: Beijing Forestry University
Advisers: Wang Xiangrong, Zhao Jing
Participants: Zhang Zhiqing, Liu Qi, Ma Lin, Wu Yanxi, Wu Yuxuan, Jiang Xin, Li Jianzhe, Huo Da

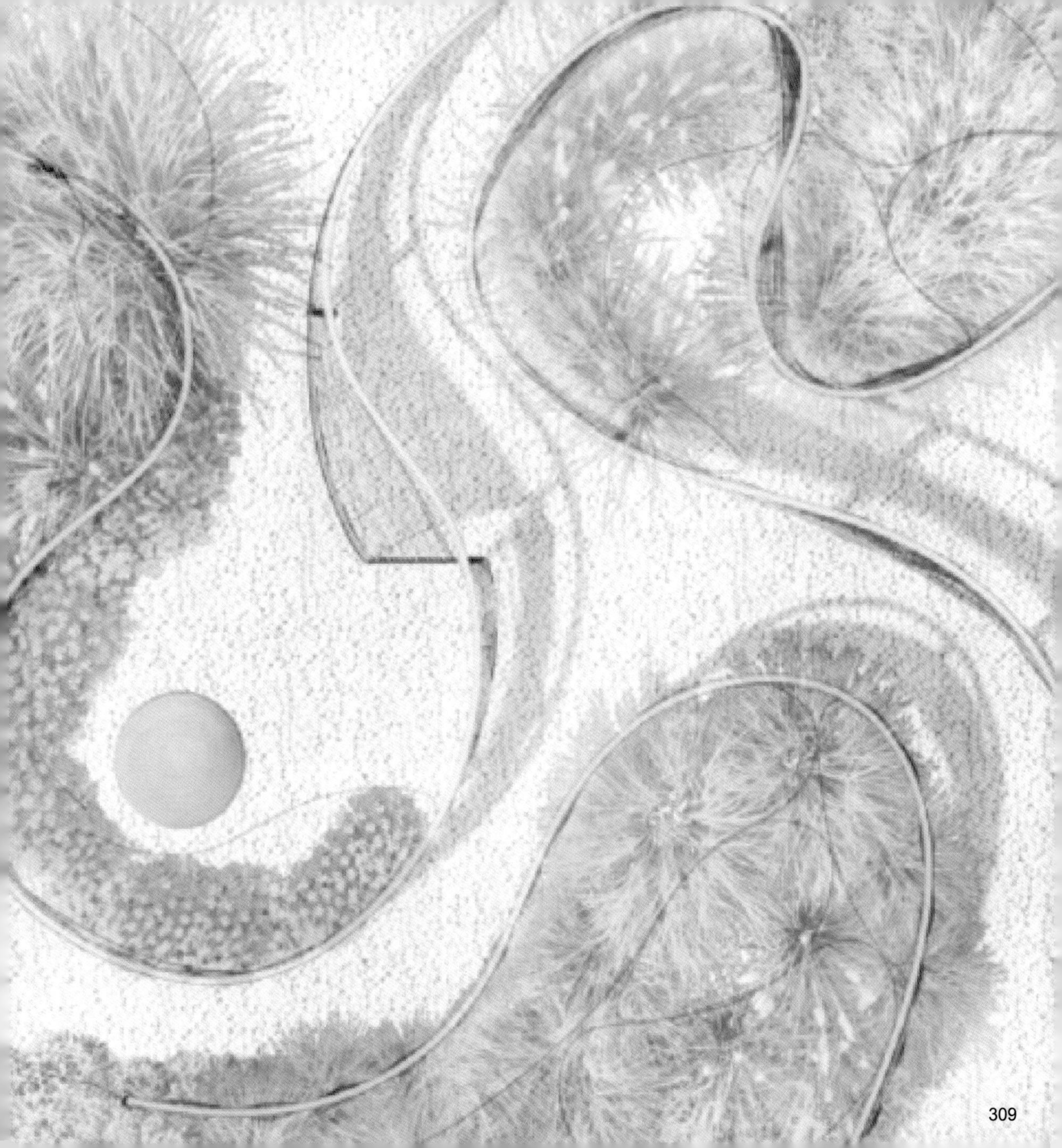

隧·燧
THE TUNNEL

隧道，是人类用勇敢和智慧，穿越险阻，通达“秘境”的利器。从“仿佛若有光”的天然峡隙，到“天堑变通途”的工程壮举，隧道沟通了无数人类曾经无法涉足的秘境，将文明的光华，传向探索的彼岸。

燧火相传，闪耀在隧道两端的，是奋进求索的光芒，是力行求仁的温暖，是秘境之美的印痕。

竹构仿佛山形的轮廓，光影斑驳的隧洞，中心“盆地”蓦然回首间的空间转换，隐喻着以成都为起点，通达秘境的燧火传递之隧，是一曲献给奋斗者的礼赞。

Tunnel is a weapon for human to access to "secrets". From the natural gorges "as if there is light", to the engineering achievements "natural gully becomes unobstructed access road", the tunnel communicates countless secret realms that humans once could not set foot in, and spreads the brilliance of civilization to the other shore of exploration.

The fire, shining at both ends of the tunnel, is the light of striving and seeking, and the warmth of striving for benevolence and the mark of the beauty of the secret.

The bamboo structure resembles the outline of a mountain, the tunnel is mottled with light and shadow, and the central "basin" reflects the spatial transformation between suddenly looking back. It metaphors the tunnel of flint-fire passing through the secret realm from Chengdu as a starting point. It is a tribute to the strugglers.

参赛者学校：西南交通大学
指导老师：周斯翔、吴然
参赛者：贺肖淇 陈镜伊 吴天昊 张润旋 黄浩鹏 唐雨倩 张柏森 刘百川
School: Southwest Jiaotong University
Advisers: Zhou Sixiang, Wu Ran
Participants: He Xiaoqi, Chen Jingyi, Wu Tianhao, Zhang Runxuan, Huang Haopeng, Tang Yuqian, Zhang Baisen, Liu Baichuan

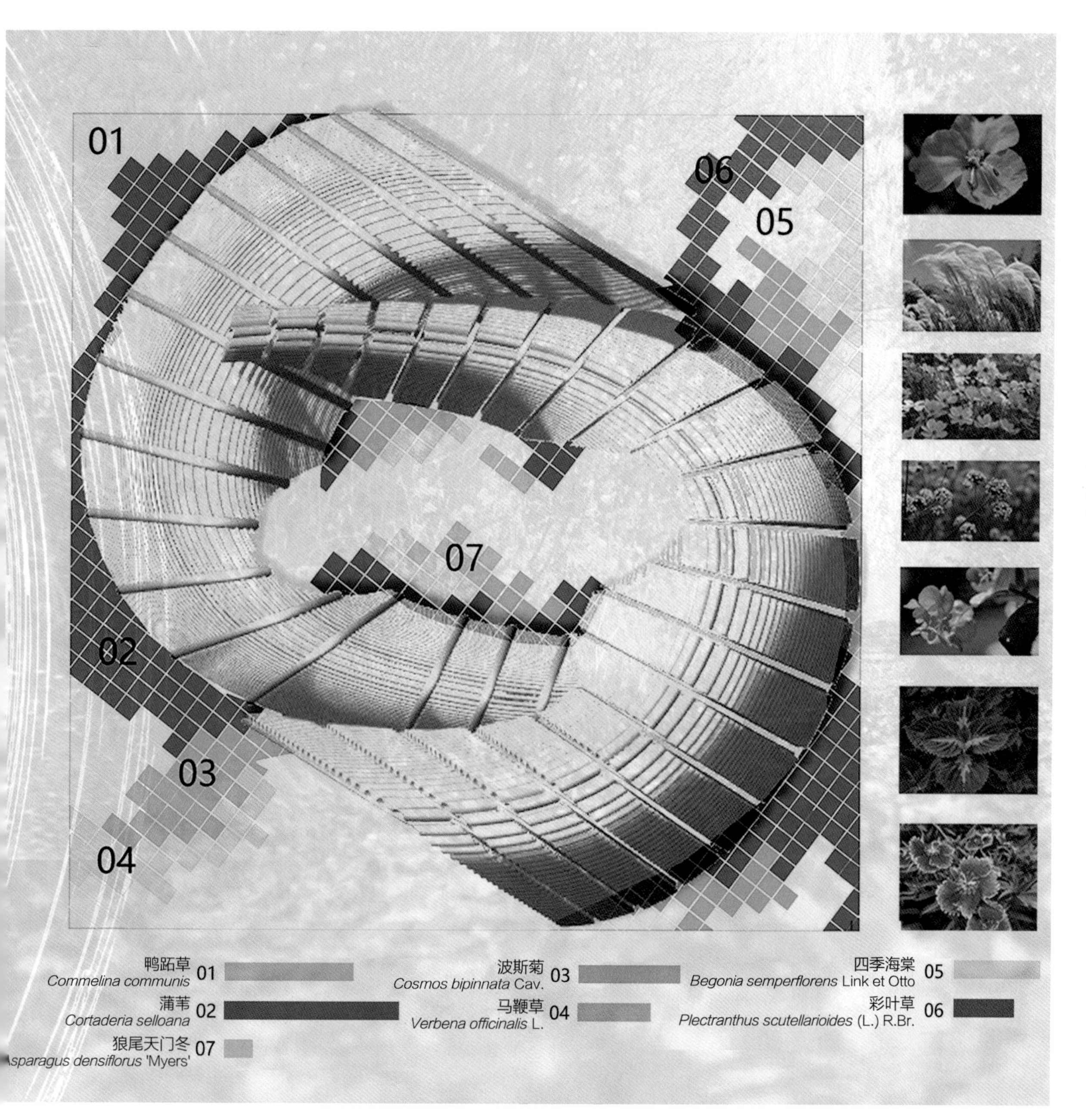
01
02
03
04
05
06
07
鸭跖草 01
Commelina communis
蒲苇 02
Cortaderia selloana
狼尾天门冬 07
Asparagus densiflorus 'Myers'
波斯菊 03
Cosmos bipinnata Cav.
马鞭草 04
Verbena officinalis L.
四季海棠 05
Begonia semperflorens Link et Otto
彩叶草 06
Plectranthus scutellarioides (L.) R.Br.

University

粉黛乱子

莫比乌斯

∞ MOBIUS BAND

设计源于拥抱这种最能体现人类情感的互动方式，并由此衍生出类似莫比乌斯环的连续性设计形式。

作为一个整体，循环的网状编织肌理在视觉和空间上无限延展，它将外部环境向内部空间引导，由外而内，自下而上，最终再次流转回外部环境，再现了静谧氛围下紧密环抱的空间感受。

半透光的竹篾表皮与内部空间和谐而统一。在其最中心，横向外部编织在空间上流动成纵向上的核心筒，自然光线得以削弱和虚化，交叠的光与影相互缠绵，形成了柔和惬意的内部空间。目光徜徉其间，在洒下的天光中，感悟真我。内与外连接的悠扬，光与影的轻盈缥缈，“秘境”便在这似透似露之间诞生了。

This structure generates from a hug — an interaction that reflects humankind emotions best — deriving the consistent form which is similar to a Möbius band.

As an organic whole, the Möbius band like contexture extends both visually and spatially. The surface guides the exterior through the interior, then flows back to the external environment. It develops the spatial sensation of being securely surrounded among tranquil surroundings.

The translucent bamboo strip skin is harmonious and unified with the internal space. In the center, the horizontal skin flows inside and becomes the vertically structural core which dims natural light intensity. Shadows are projected and overlapped to create a cozy internal space that resembles embracement in appearance. Visitors could look up and down to experience the space in sunlight. With the melodious connection between interior and exterior, a "secret space" is created within the refraction of lights.

参赛者学校：北京建筑大学

指导老师：张振威、李梦一欣

参赛者：韩希元、贾雨薇、刘丹帅、陈肇晖、刘雨杉

School: Beijing University of Civil Engineering and Architecture

Advisers: Zhang Zhenwei, Li Mengyixin

Participants: Han Xiyuan, Jia Yuwei, Liu Danshuai, Chen Zhaohui, Liu Yushan

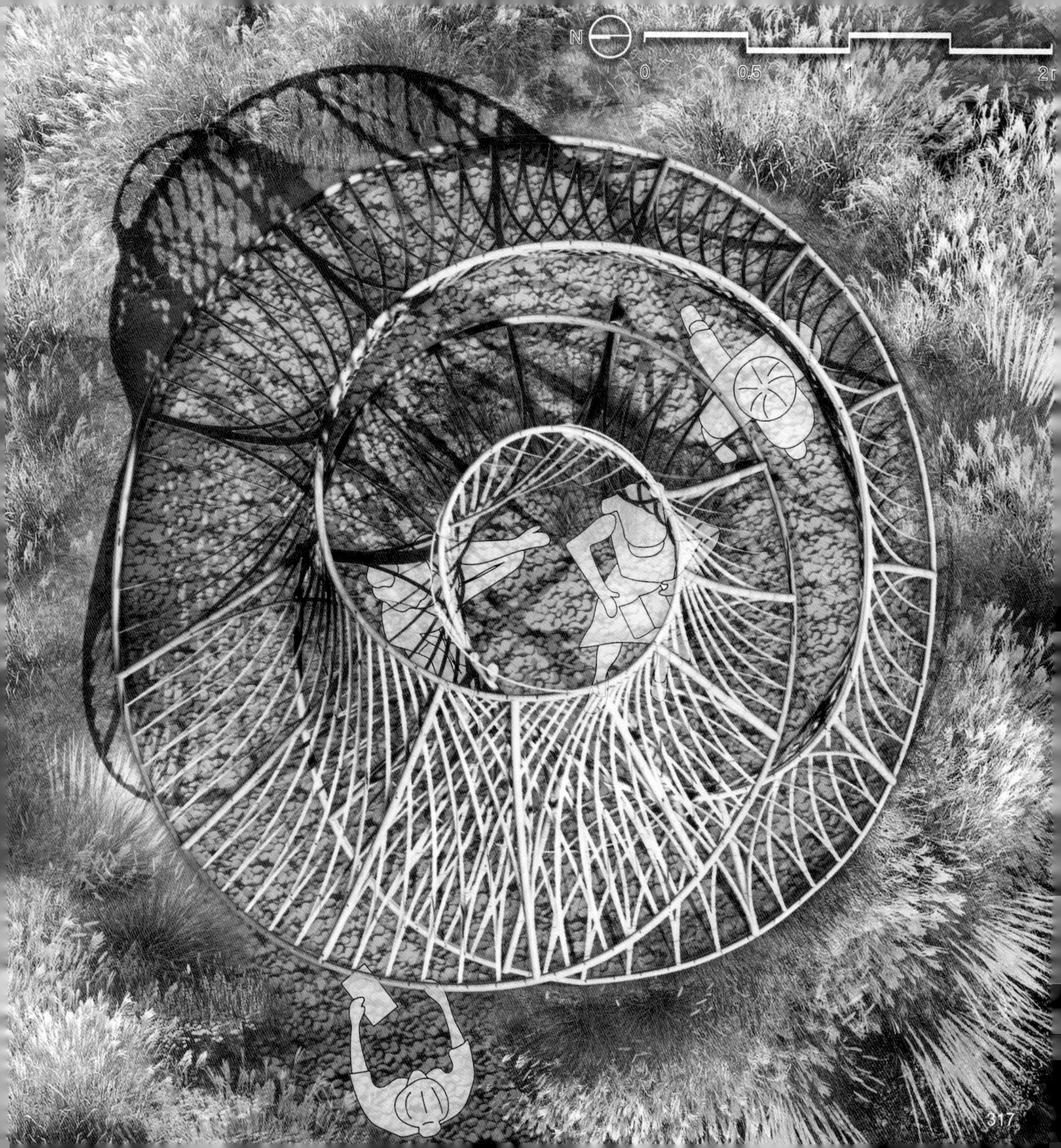
N
0
0.5
1
2

莫比乌斯
北京建筑大学
协作单位
鼠尾草

紫叶酢浆草
Oxalis triangularis 'Urpurea'
酢浆草科 酢浆草属
狐尾天门冬
Asparagus densiflorus 'Myers'
百合科 天门冬属

月中梦蝶
DREANING BUTTERFLY UNDER THE MOON

人们向往蝴蝶的自由，却如庄周一般被困于现实，但依旧不放弃在现实生活中追逐梦境的美好，追求自由的境界，梦境与现实交织。

蝴蝶与人在梦境中相互转化，现实与梦境的转化，虚与实的转化，在转化的过程中，就像是内心走不出的轮回，便不知是庄周蝶梦还是蝶梦庄周，最后达到物我两忘的境界，这便是秘境。

设计采用月洞门元素，作为现实与梦境转化的介质，穿梭在月洞门里，犹如穿梭在梦之秘境中。

People yearn for the freedom of butterflies, but like Zhuang Zhou, they are trapped in reality, but they still do not give up chasing the beauty of dreams in real life, pursuing the realm of freedom, and dreams are intertwined with reality.

The transformations between butterflies and human beings, reality and dreams, virtual and real are like a reincarnation that cannot be walked out of the heart. It is not known whether it is Zhuang Zhou dreamed of a butterfly or a butterfly dreamed of Zhuang Zhou. The realm of things and I have been forgotten, this is the secret realm.

The design adopts the moon cave gate element as a medium for the transformation of reality and dreams. People travels through the moon cave gate as if they are in the secret world of dreams.

参赛者学校：西南民族大学
指导老师：曾昭君、陈娟
参赛者：田琳璐、刘喻、冯志军、于晨曦、江钰栎、李荣乔、李航
School: Southwest Minzu University
Advisers: Zeng Zhaojun, Chen Juan
Participants: Tian Linlu, Liu Yu, Feng Zhijun, Yu Chenxi, Jiang Yuli, Li Rongqiao, Li Hang

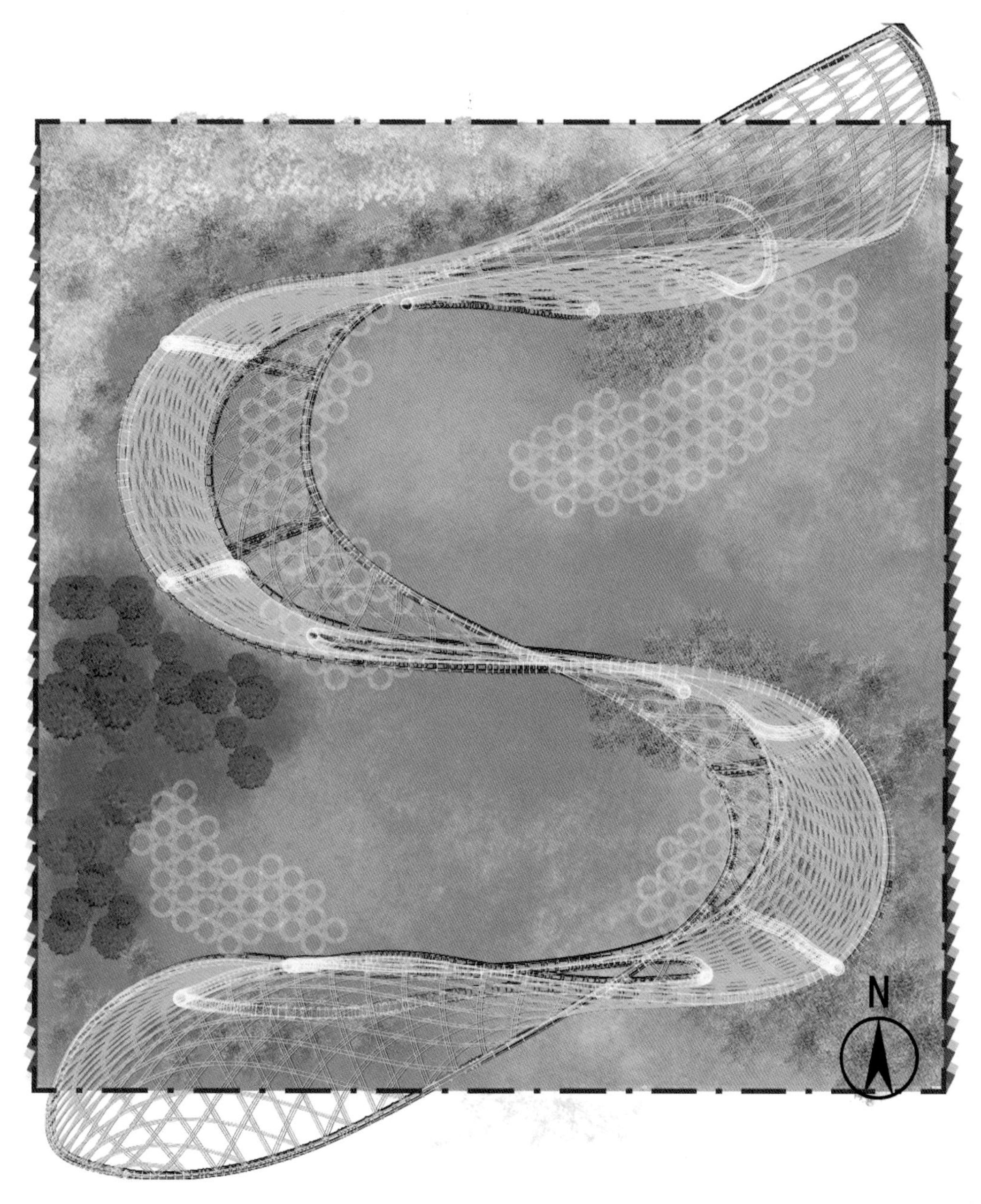
N

成都国际花园节
作品建造中

织影

WEAVING SHADOW

世间的光影本是转瞬即逝的，但若用实体的竹子，通过竹条的弯曲与编织，便可让编织光影成为可能。首先，通过平面的路径编织，营造曲径通幽、探寻理想之感。整个平面既像是一个迷宫，又像是一个游园，人们也许会在其中迷失，也许会寻找到自己的理想乐园；然后通过立面编织，营造光影重重，若隐若现之感；最后通过空间的编织营造山重水复、柳暗花明之感。通过平面、立面及空间的共同编织，最终构成了竹径花园，也构成了自己的理想。

编织光影的步骤共分为三步：首先，通过平面的编织——鸿渺浮云低：平面上如同迷宫般路径的变化如同空中浮云，充满了多样的变化。然后通过立面的编织—— 雁鸣秋思远：立面上不同条状编织的交错像空中秋雁飞过的痕迹，充满动感。最后通过空间的编织——穿帘风入竹：空间上的整体编织如同风吹过片片竹帘，通过变幻角度及视窗，给人以全新的空间感受。

Light and shadow in the world are fleeting, but if we use physical bamboo, through the bending and weaving of bamboo strips, weaving light and shadow can become possible. We first weave through a flat path to create a serene path and a sense of searching for ideals. The whole plan is like a labyrinth and an amusement park. People may get lost in it, or find their ideal paradise; Afterwards, through the weaving of space, a sense of multiple landscapes and bright willows is created. Through the joint weaving of plane, facade and space, we finally constitute our bamboo garden, and also constitute our own ideal.

The steps of weaving light and shadow are divided into three steps: First, the weaving of the plan – the swan flying between the low clouds: the change of the path like a maze on the plan is like the clouds in the sky, full of various changes. Secondly, the weaving of the facade – the call of the geese renders the artistic conception of autumn: the different strips weave on the facade are staggered like the traces of autumn geese flying in the sky, full of movement. Finally, the weaving of space – through curtains and wind into bamboo: the overall weaving in the space is like the wind blowing through pieces of bamboo curtains, through changing angles and windows, giving people a brand–new spatial feeling.

参赛者学校：北京林业大学
指导老师：郑小东、赵晶
方案阶段参与者：辛昊儒、邱天琦、张敔瑞、胡坤宁、余孟韩、吴雅轩、倪凌珊、刘育军
建造阶段参与者：练则可、陈然、沈子晗、郝慧超、姜昕怿、邵壮、张烨、凌怡晨
School: Beijing Forestry University
Advisers: Zheng Xiaodong, Zhao Jing
Program Stage Participants: Xin Haoru, Qiu Tianqi, Zhang Qirui, Hu Kunning, Yu Menghan, Wu Yaxuan, Ni lingshan, Liu Yujun
Construction Stage Participants: Lian Zeke, Chen Ran, Sheng ZIhan, Hao Huichao, Jiang Xinyi, Shao Zhuang, Zhang Ye, Ling Yichen

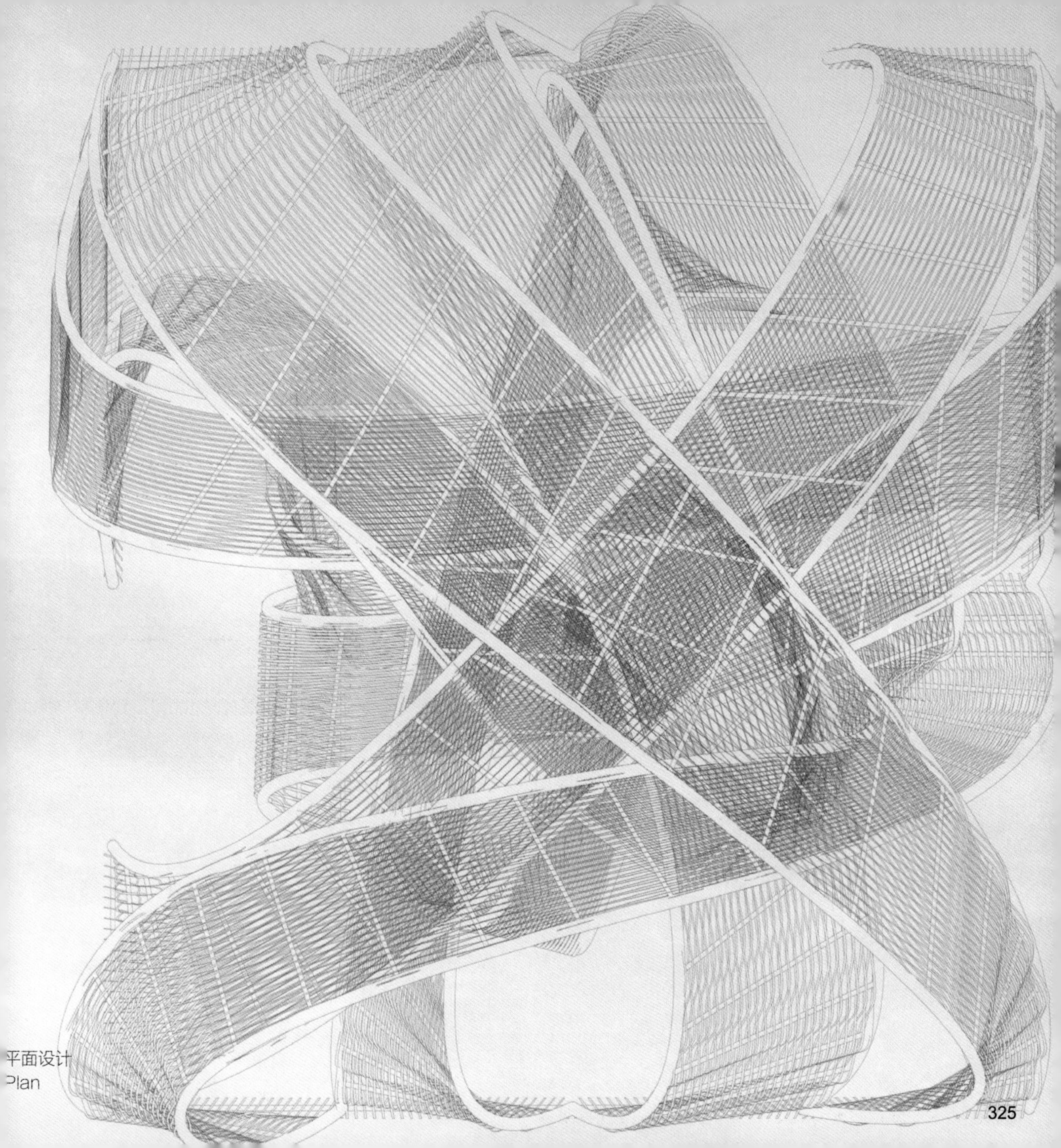

平面设计
Plan

矮蒲苇

花叶络石

蜀山秘境
FAIRY IN SHU

“天人合一，道法自然”是人与自然共生的最高境界，夫道者，万物之源也，衍而万化，是为自然；心者，思辨之源也，拓以八方乃识道也，是以寻道之路亦是追寻物我互化之路，上下求索，方能于自然之境中有所感悟。本作品以“天人合一”的道家思想为灵感，利用竹的弯曲、螺旋塑造蜀地自然风貌，构建“入、寻、悟、离”四重空间体验，使人在穿行于竹境的过程中能感受到人与自然和谐共生的哲学境界。

The unity of man and nature is the highest level of coexistence between man and nature. "Tao", the root of all things, evolves into all things, which is nature. The heart of man is the source of speculation and extends to the principle of eight talents. Therefore, the way to seek Tao is to pursue the road of mutual transformation of the object and the self. Only by searching up and down can we gain insight in the natural environment. Inspired by the Taoist thought of "unity of man and nature", this work uses the bending and spiral of bamboo to shape the natural landscape of Shushan, and constructs the four-dimensional experience of "entering, seeking, enlightening, and leaving", so that people can feel the philosophical realm of accepting the harmonious coexistence of human and nature in the process of walking through the bamboo environment.

参赛者学校：西南大学
指导老师：周建华、孙松林
参赛者：邓思桥、李武肸、李乾、丁双鑫、侯文武、祝赛男、白余迦、张梦屿
School: Southwest University
Advisers: Zhou Jianhua、Sun Songlin
Participants: Deng Siqiao, Li Wuxi, Li Qian, Ding Shuangxin, Hou Wenwu, Zhu Sainan, Bai Yujia, Zhang Mengyu

蜀中茶器
THE TEA SET OF BASHU

斗转星移，沧海桑田，不变的是巴蜀人民寄于汤茶之上的闲逸生活态度。而茶文化，也作为巴蜀文化中的典型代表符号之一，在这片土地上生根、滋长。

设计以巴蜀茶文化为引线，以时间为长轴，将古今印象复合于一个抽象竹制构筑物内，以此体现人地共生的文明发展脉络。

Time has changed, but what remains the same is the relaxed attitude of Bashu people to life on tea. Tea culture, as one of the typical symbols of Bashu culture, took root and grew in this land.

The design takes the tea culture of Bashu as the lead line and time as the long axis. It combines the ancient and modern impressions in an abstract bamboo structure, so as to reflect the development vein of civilization of symbiosis between man and earth.

参赛者学校：四川大学
指导老师：李沄璋、李恒
参赛者：吴颖、孙禄鹏、王瑾瑄、张慧珍
School: Sichuan University
Advisers: Li Yunzhang, Li Heng
Participants: Wu Ying, Sun Lupeng, Wang Jinxuan, Zhang Huizhen

矮生五色梅

悬镜

THE MYSTERY OF WONDERLAND

拱弧，以作门之象征，寓意通往秘境的通道。上下相应喻天境地境，仿佛此身此世与仙境朦胧。竹直而列与竹弧而错形成对比，其间错落镜像，时空交与一处，自成沙漏状，静默时间之流。界，若竹中秘境，缥缈相隔。

秘境亦为心境，出世入世本无界，鱼之乐快然自得。框架相引，渐入佳境。边无所界，故以竹编蒙之，取其虚无缥缈之境。席地而坐，俯仰天地，若庄生晓梦忘却时空，不知今夕何夕。

Arch is the symbol of the door, representing the passage to the secret places. Top and bottom represent heaven and earth respectively. It makes us feel as if in the Pure Land. Bamboo poles can be arranged in straight lines or curved, forming a contrast between each other, like images in a mirror. Time and space converge here, naturally forming the shape of an hourglass that represents the time goes by, like a secret garden in a bamboo forest that can separate us from the other world.

The real secret place is a simple place in everyone's heart, actually there is no clear boundary. The frame serves as a sign of guidance, and the internal is like a fish swimming happily, attracting people to enter. The boundary is not completely restricted. It is made of bamboo composition which is masked into mat, creating an ethereal and distant artistic conception. Sitting on the ground, looking at the sky and the earth, like a dream, where people can forget the time, and don't know what is true.

参赛者学校：山东建筑大学
指导老师：宋凤、李学东
参赛者：苗行健、庞国航、孙姿旋、王炜、姚海啸、刘子墨、王宇鹏
School: Shandong Jianzhu University
Advisers: Song Feng, Li Xuedong
Participants: Miao Xingjian, Pang Guohang, Sun Zixuan, Wang Wei, Yao Haixiao, Liu Zimo, Wang Yupeng

方案生成一
Scheme generation I:
拱一门
Arch-Entrance
生成空间
Space formation
镜像空间
Space Mirroring
悬空框体
Hanging frame
支撑斗拱
Arch supporting
方案生成二
Scheme generation II:
沙漏拉线
Hourglass cable
花境赋土
Garden covering
侧拱纺织
Side arch weaving
界面围合
Space enclosure
底面纺织
Underside weaving
此
Front
彼
Profile
天
Top
地
Bottom
时
Time
空
Space

悬 镜

太虚幻境
ILLUSORY LAND OF GREAT VOID

方案受《红楼梦》中太虚幻境的启发所成，结合仙女元素，融入山云形态，以花香与舞蹈搭建整个秘密花园。

通过两种不同的编织方式，将空间划分为虚实相间的两部分，人可在竹编中穿梭，也可在内部空间席地而坐，顶部开口，亦可在静谧的夜晚仰望星空。

在繁忙生活中，会有这样一处栖居之地，倾听你的遐想，守护你的秘密。

The plan was inspired by the fantasy of the *Dream of Red Mansions*, combined with fairy elements, integrated into the form of mountains and clouds, and built the entire secret garden with flowers and dance.

The plan divides the space into two parts, the virtual and the real, through two different weaving methods. People can shuttle through the bamboo weaving, they can also sit on the floor in the internal space, the top is open, or they can look up at the starry sky in a quiet night.

In a busy life, there will be such a place to live, listen to your reverie, and protect your secrets.

参赛者学校：西北农林科技大学
指导老师：张新果 罗西子
参赛者：蒲宝婧、刘恒君、陈江为、龙科良、肖旎珺、王皓松、刘熙宇、杨雪刚
School: Northwest A & F University
Advisers: Zhang Xinguo, Luo Xizi
Participants: Pu Baojing, Liu Hengjun, Chen Jiangwei, Long Keliang, Xiao Zhaojun, Wang Haosong, Liu Xiyu, Yang Xuegang

太虚幻境

相关活动与报道
Related Activities and Reports

开放展览 Open Exhibition

活动 ACTIVITIES

开放展览 Open Exhibition

公园城市

公园城市 花重锦城

“最美花园”线上投票
Online Voting for "The Most Beautiful Garden"

活动 ACTIVITIES

现场手作教学 On-site Construction Teaching

活动 ACTIVITIES

摄影作品征集 Call for Photography Works

赏成都
迎大运
AWARENESS
92
Be a voice,
not an echo!
Instagram
COMPASSION
23
2020年成都
公园城市国际花园节
暨第三届北林国际花园建造节
暨第二届英国体育节
暨成都市第三届健身瑜伽展示交流公开赛
THIRD FITNESS YOGA EXHIBITION AND EXCHANGE OPEN COMPETITION
2020年11月15日
主办单位：成都市瑞
KELME

活动 ACTIVITIES

悦动成渝 解锁别样体育节

Vigorous Chengdu and Chongqing: Unlock the Special Sports Festival

生活美学 共享时尚生活展
Life Aesthetics: Fashion Sharing Exhibition

公园城市　花重锦城

2020 成都公园城 … 花园节暨第三届北林国际花园建造节

RE
ADA
2019 世界风景园林师高峰讲坛
2019 International Landscape Architects Symposium
2019
10/12–10/13
RESILI
SUSTA
RANS
2019 International

活动 ACTIVITIES

同期论坛讲座
Concurrent Forum Lecture

活动 ACTIVITIES

亲子活动 Family Activities

每年都会在花园节的后续开展一系列开放活动，亲子活动就是其中之一。活动借助设计作品，为家长和孩子们提供丰富的游园体验。

Every year, a series of open activities would be carried out after the garden festival, and parent-child activities were one of them. The design works could provide parents and children with a rich garden experience.

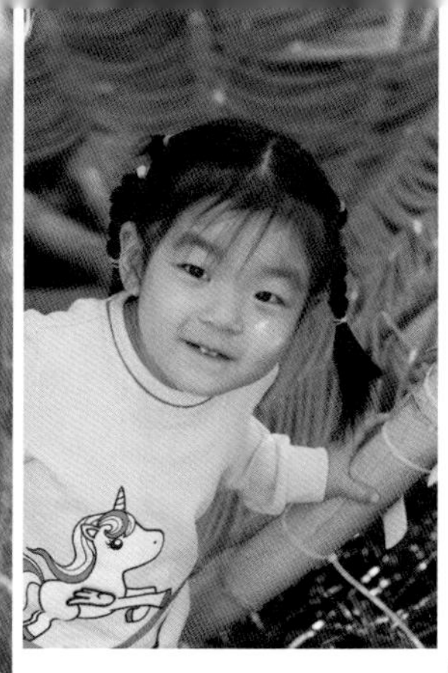

学习创新能力是学生综合素养的重要体现，在日前举办的第二届北京林业大学国际花园建造节上，中外大学生的园林设计创意吸引了众多目光。

材料建构和艺术表达相互交叠运用的建造体验，激发了园林学子的创作热情，提高了动手能力，弘扬了工匠精神，为风景园林行业培养实践创新人才。

这是北京林业大学推进世界一流学科建设，为风景园林行业培养实践创新人才的重要举措。

将中华传统诗词之美与建造艺术之美紧密相连，竹构花园绽放北京林业大学校园。

Innovation ability is an important manifestation of students' comprehensive quality. At the second BFU International Garden-Making Festival held a few days ago, the garden design creativity of college students from home and abroad attracted a lot of attention.

The construction experience of material construction and artistic expression overlapping each other has stimulated the creative enthusiasm of students, improved operational ability, promoted the spirit of craftsmanship, and cultivated practical and innovative talents for the landscape architecture industry.

This is an important measure taken by BFU to promote the construction of world-class disciplines and cultivate practical and innovative talents in the landscape architecture industry.

Closely linking the beauty of traditional Chinese poetry with the beauty of construction art, bamboo gardens are blooming in the campus of BFU.

《以诗为序，以花衬竹 | 15 个竹构花园今日绽放北林大校园》——北京林业大学

《竹构花园 · 诗意殿堂丨 2019 第二届北林国际花园建造节开幕》——北林园林资讯

《花园的诗意丨 2019 第二届“北林国际花园建造节”主题发布及设计竞赛方案征集》——风景园林杂志

《北林国际花园建造节的 30 个竹构作品，有意思！》——景观邦

gooood

《大风起兮云飞扬 — 云在亭 · 北京林业大学花园节信息亭 / 素朴建筑工作室》——谷德设计网

《北京林业大学花园节信息亭“云在亭” / 素朴建筑工作室》——建日筑闻

《竹境丨 2019 第二届北林国际花园建造节成功举办！》——世界竹藤通讯

《【适之】竹构花园 · 诗意殿堂丨 2019 第二届北林国际花园建造节开幕》——园艺疗法

《魅力竹建筑丨“竹构花园”绽放第二届北林国际花园建造节》——中国竹产业协会

《多种类型亭廊设计》——景观资源小能手

"Preface with Poems, Presenting Bamboo with Flowers I 15 Bamboo Gardens Blooming Today in BFU Campus" — "Beijing Forestry University"

"Bamboo Garden·Poetic Palace | The Opening of the 2nd BFU International Garden-Making Festival in 2019" — "LA News of BFU"

"The Poetic of Garden | The 2nd "BFU International Garden-Making Festival" Theme Release and Design Competition Calling for Entries" — "Landscape Architecture Journal"

"The 30 bamboo works on BFU International Garden-Making Festival are interesting! " — "Landscape Union"

"The wind is blowing and the clouds are fluttering—Swirling Cloud Pavilion, Beijing Forestry University Garden-Making Festival Information Pavilion / SUP Atelier" — "Gooood Design"

"BFU Garden-Making Festival Information Pavilion 'Swirling Cloud Pavilion' / SUP Atelier" — "ArchCollage"

"Bamboo Landscape | 2019 2nd BFU International Garden-Making Festival was successfully held! " — World Bamboo and Rattan Magazine

"Bamboo Garden·Poetic Palace | The Opening of the BFU International Garden-Making Festival" — "Horticultural Therapy"

"Charming Bamboo Architecture | "Bamboo Garden" Blooms for the 2nd BFU International Garden-Making Festival — "China Bamboo Industry Association"

"Various Types of Pavilion Designs" — "Landscape Resources Assistant"